EQUILIBRA

The Philosophy and Political Economy of Existence and Extinction

DR. PAUL OQUIST

Printed in the United States of America.

978-0-578-66989-2 – (Hardcover)
978-0-578-67193-2 – (Paperback)
Also available for Kindle and eBook readers.

Dedication

EQUILIBRA *is dedicated to Father Miguel d'Escoto (RIP), whose final message was the urgent spiritual need for a Survival Movement for Humanity, other species, and Mother Earth.*

It is also dedicated to Edith West Fell (RIP), Harry Fell (RIP), Barbara Edith Kelley (RIP), Paul Herbert Oquist, Sr. ("Big Swede") (RIP), Harry David Oquist (RIP), Pilar Aguilar Matamoro de Oquist, and Patricio Oquist Aguilar, to all of whom I am grateful for the myriad support received in this 20-year effort.

Acknowledgements

*From President Daniel Ortega Saavedra
and Vice President Rosario Murillo Zambrana,
I have received both wisdom and inspiration.*

*This volume was finished in the instant tranquility
of the Venado tree house, overlooking Redonda Bay,
Aqua Resort, Tola, Rivas, Nicaragua.*

The author wishes to express his appreciation for the multiple contributions of Daniyal Aziz. Morten Nygart has given me by example the patient perseverance needed in large, long-term projects. Scarlett Nygart has shown me by example the impatient drive needed to get things done day after day. Alexander Nygart opened for me the fascinating world of the technological revolution in social communication. The author is grateful for the assistance in editing this work by Kate Reynolds and finalizing the text from Maria Fernanda Gutiérrez Gaitán.

CONTENTS

PREFACE

EQUILIBRA CONSISTS OF:

The Ultimatum and the Philosophy of Existence and Extinction
The Political Economy of Existence and Extinction

Each **EQUILIBRA** volume consists of themes formed by 10 numbered points of 3 or 4 lines. This degree of conciseness is what allows covering the vast conceptual territory of existence and extinction in manageable form.

The points are numbered because **EQUILIBRIA** is designed to be a collective, interactive, evolving, live work. Readers are invited to correct errors of fact and improve upon interpretations and conclusions by sending proposed replacement points (identified by their numbers), or of an entire theme of points, to the author at:

Replacements1@EQUILIBRA

The author will name and chair an **EQUILIBRA** Panel that will decide what changes should be made to update, correct, and improve the second edition of each Volume. Contributors will receive footnote credits for their enrichment of the positions. This should be a contribution of a concerned community to a Survival Movement.

Additionally, a website will be available where discussion groups can be formed on different themes and/or individual points. These conversations can also be considered by the EQUILIBRA Panel for enrichment of the text.

EQUILIBRA is based on philosophy and the history of the Universe, Planet Earth, the emergence and evolution of life on Planet Earth, including homo sapiens sapiens, of political economy, as well as on contemporary policy analysis. It strives to be based on science to the maximum extent possible, but it is not science. At best it is informed by science. When an acceptable scientific conclusion is not available, EQUILIBRA takes the best knowledge available to generate intermediate and common-sense conclusions. This is not a compromise but rather the way our knowledge works in every sphere. Only a small portion of our knowledge is scientific, and this generally comes combined with other forms of knowledge.

The philosophical positions of EQUILIBRA are based on the study of structured, conditioned transformations within patterns of life cycles. From these are derived the logic and principles of the Universe, evidence of which we access through interrelated sensory induction and cognitive deduction.

At very least these are my positions, what I believe in 2020. With diffusion, the interactive process, and the discussion groups, these positions can become both enriched and held by a broader base in a Survival Movement.

VOLUME I

The Ultimatum and the Philosophy of Existence and Extinction

Executive Summary

THE EQUILIBRA SERIES BEGINS with THE ULTIMATUM, the myriad ways Mother Earth is communicating to us the current, imminent, and potential **desequilibra** that threaten the ecosystems and the synergies between them that support life on planet Earth. The ULTIMATUM is imperative—transform ourselves now or decline and perish as a result of the consequences of our own actions, a species suicide. The ULTIMATUM is dual: we can produce our own extinction and that of other species through weapons of mass destruction, especially nuclear, and/or environmental degradation, especially climate change.

Our pathetic response and ever increasing transformative and destructive capacities are also described, as well as their impacts and our societal incapacity to deal with them. It was once believed that the Earth was the center of the Universe, but in practice we still operate as if human beings are the center of the Universe. Species centrism considers us to be the owners of the Earth and all of its resources, which we can dispose of at will, no matter what the consequences.

Endless, limitless, mindless growth, production, consumption, and capital accumulation cannot continue on a planet with limited, degraded resources. The formula that has allowed us to survive and advance, transforming nature, is now counterproductive, making fundamental corrective action most difficult. It is easier for us to continue transforming nature than to transform ourselves.

This is illustrated by the nine alienations from reality that are leading us to extinction. They are eternalism, supernaturalism, developmentalism, hyperindividualism, immediatism, capitalism and the highly interrelated financialism, scientificism, and fatalism with its two variants, providentialism and nihilism.

The depth of our dilemma is that the hegemonic economic and political structures and their media are largely based on short-term gain despite long-term consequences. The monopolists, bankers, and speculators cannot be controlled by the politicians, because the monopolists, bankers, and speculators control the politicians.

Basic social change in recent centuries has come about through broad-based international social movements of people in different countries and sectors, employing different tactics, undertaking different activities, but all working for a common definable goal with historical persistence (e.g., abolition of slavery, women's suffrage, 8-hour working day, independence from colonialism).

We have lost touch with our place in the evolution of the Universe and Mother Earth. Our most common answers to the big questions have worn thin and lack the spiritual urgency that can be the basis for a Survival Movement. For this we need to have an understanding of the Universe and Mother Earth and how we are an integral part of evolution, collectively and individually, that renews our identity and values, and elevates our level of consciousness of our current situation and the vulnerabilities and risks that threaten our existence. What is required is transformation through a Survival Social Movement that constructs a low-carbon, climate-resistant and resilient, circular, sustainable society with far greater equality.

Box 1: The Big Questions

Life is always a struggle for species and individual survival, even at the physiological level of reflexes, fight/flight, and homeostatic equilibria. The meaning of life is to live it, hopefully to the fullest, but at least well, with regard to safety, wellness, and well-being derived from survival, as well as whatever purposes we can construct beyond that to add quality to our existence. Others judge our level of success in these pursuits; only we can assess the level of satisfaction.

Where we come from is the expansion of 13.77 billion years of Universe. The Genus Homo first appeared 2.6 mya (million years ago), and *Homo Sapiens* evolved from 300,000 to 200,000 years ago in Africa, from where we spread to the rest of the world except Antarctica.

Where we are is in the Supercluster Virgo, subgroup Local of over 30 galaxies, the largest being Triangulum, Andromeda and our own galaxy the Milky Way (one of the oldest at 3.6 bys), where our Sun and the Solar System are located in the Orion Arm, some 25,000 light years from the center of the galaxy. We are the third rocky planet from the Sun at 147.18 million kms., Mercury and Venus being closer and Mars the next.

The scientific answer as to who we are is the sub-species *Homo Sapiens Sapiens*, species *Homo Sapiens*, genus *Homo*, family Hominidae, suborder Haplorhini, order Primates, class Mammalia, phylum Chordata, Kingdom Animalia, an integral part of life on the planet Earth.

The poetic answer is that we are offspring of Grandfather Singularity, Grandmother Galaxies, Father Son, Mother Earth, an integral, natural part of the logic, principles, and evolution of the Universe, star dust to be sure.

The two volumes address the need for understanding and action through the Ultimatum and the Philosophy and the Political Economy of Existence and Extinction.

The Philosophy of Existence and Extinction is presented through a metaphysics, an ontology, an epistemology, and an ethics.

The *metaphysics* theorizes on the Multiverse and the Singularity and their characteristics. Also described are the logic and principles of the Universe. The logic is one of conditioned, structured transformations within life cycles, not randomness, but with a pinch of chance. The principles are **EQUILIBRA/desequilibra** within the unity of opposites, the change and simplicity to complexity constants, identity in diversity and information that is posited to be the theory of everything. (See Box 2.)

BOX 2: Information as the Theory of Everything

Information is contained in and structures the Universe. The logic and principles of our Universe construct and deconstruct entities and processes through forces, energy, and matter in time within their life cycles as per information.

The "cosmic stem cell" of the Singularity contained all the logic, principles, energy, forces, matter, and energy of the Universe as information. In the Great Expansion "Cosmic DNA information" structured evolution through the definition of potentialities of entities and their evolution or not, as per conditions. At reducing temperatures/speeds/densities, there was a decoupling of energy, forces, and different forms of matter, followed by combinations, both as per information.

Different types of stars are constructed and destructed according to the information of the life cycle of that type of star. When a star is swallowed by the gravitational attraction of a black hole singularity, it leaves behind information of its previous existence.

Organic molecules read other organic molecules and react accordingly. Our immune system reads the markers of pathogens and reacts accordingly. Genetic inheritance information is stored in DNA and communicated by RNA and proteins to cells in the process of formation of living organisms. Human beings are conceived when the information of 23 male chromosomes is combined with the information of 23 female chromosomes. Life begins with a single cell that divides into a second, they and subsequent cells communicate, genes are turned on and off in determinate sequences, as a living being is created according to the information of its specific DNA. This is the case for humans as well as all other living beings, each with its specific DNA or RNA.

The Universe and life read, are structured by, and adapt to information, as well as maintain homeostatic **EQUILIBRA** by processing information. **The Theory of Everything is Information.**

The *ontology* consists of two parts, one on existence and the other on extinction. The *ontology of existence* describes the Great Expansion based on the best information now available, through the point in the evolution of life on planet Earth of the appearance of the species *homo sapiens sapiens*. From that point onward human development is described as the **Political Economy** of cultures (Paleolithic, Mesolithic, Neolithic, Bronze Age, Iron Age, Civilizations, Six Stages of Imperialism, Political and Technological Revolutions), described in Volume Two.

The *ontology of extinction* posits the inevitability of species extinction and individual death. All that we have is a window of space-time to exist. What we can do is avoid decreasing the time of species existence and individual survival, as well as prolong the meaning, enhance the purpose, and increase the quality of life.

The five major historical mass extinctions and their causes are described. The Sixth Holocene event is ongoing and directly linked to our impact on other species, the environment, habitat degradation, and climate change. Imminent threats include human-caused degradation of the marine biota and the declining population of the largest, wild, land mammals, as well as bees, frogs, and the smallest living things that have very fragile habitats.

Structural threats include the cosmological (meteors, quasar electromagnetic pulses, solar radiation); geological (volcanic winter); epidemiological (antibiotic resistant bacteria and rapidly mutating viruses); nutritional (soil, water degradation, plant and zoosanitary and pest threats); climatological (human-induced global warming, droughts, floods); weapons of mass destruction (thermonuclear, nuclear, electromagnetic pulse, chemical, biological); and the new generation of human-made threats: non-biodegradable refuse, especially plastics; new materials, nanotechnologies, genetic engineering, gene editing of immune system (CRISPR); and, artificial intelligence, biomechtronics, cybernetic cyborgs, biorobotics, especially robotics with autonomous learning and self-replication, with robot soldiers being the most dangerous. Like in science fiction books and films they might conclude that humans are a virus-like creature destroying life and damaging Mother Earth and decide to try to exterminate the threat.

The hegemonic economic, social, and political structures, as well as mechanization, automation, and artificial intelligence have concentrated wealth to the point where 1% of the world's population controls over 50% of the world's wealth. At the same time wealth redistributive mechanisms have been weakened. If this continues with the imminent rapid expansion of artificial intelligence, nearly worldwide economic, social, labor, and political instability and crises are completely predictable.

Decision making and self-replicating robots pose the threat of loss of human control and entering into competitive, conflictive, and dominating relations with humans. Cybercrime and cyberwarfare are already a reality, and with the further digitalization that artificial intelligence will bring, cyber domination could become the major threat with autonomous robots among the competitors.

Given that we have serious threats we cannot control, that the most probable risks are now of human origin represents a high level of species masochism. The question is whether intelligence and foresight can generate corrective action before a final, cataclysmic breakdown obliges us to act after maximum damage has already been suffered. Does anyone doubt that the species marching toward extinction because of our impact on the environment does not include our own?

The Poetic Conclusion is that we live in a cosmic shooting gallery of meteors, comets, solar radiation, electromagnetic pulses, and gamma rays, on a molten bomb of a planet under attack by bacterial and rapidly mutating viral pathogens, unable to control the consequences of our own science and technology, especially nuclear weapons, climate change, and artificial intelligence.

The Novel Coronavirus and COVID-19 demonstrate how unprepared we are to confront our high risk threats and how little we invest politically and economically in avoiding them, which also applies to the threats of nuclear weapons and climate change. The degree of disruption of our lives, societies, and economies of the Novel Coronavirus and the disease COVID-19 is greater than 9/11, the 2007-2009 financial crisis and subsequent Great Recession combined. Yet, the disruption of the Coronavirus is small, transient, and recoverable, compared to the total, permanent, and irreversible damage of a nuclear exchange or the disruption of the eco-systems and synergies between them that maintain life on Plant Earth, including the critical variable temperature.

The **Epistemology of Existence and Extinction** posits that we perceive/process data/produce information because the Universe consists of structured, conditioned transformations, within life cycles. There are ***four interrelated forms of perception/information:*** **physiological/neurological, impulse/sub-conscious, emotion/mood,** and **conscious thought**, which contains ***five interrelated components of knowledge production***: description, explanation, prediction, preference, obligation. There are ***three levels of knowledge production*** based on level of consciousness and methodological rigor: common sense/intuitive, intermediate/ideological, and formal/scientific. Most of our daily activity is based on common-sense knowledge. This is culturally conditioned to a large extent. (See Box 3.)

BOX 3: The Cultural Conditioning of Conscious Thought

If one asks young children in the Orient and Occident about a wooden block, a rubber ball, and a metal triangle, **cultural conditioning** is apparent. Children in the Orient tend to answer, "What is this?" with wood, rubber, and metal, whereas an Occidental child tends to answer a block, a ball, and a triangle, substance vs. form.

When presented with a picture of an airport check-in area with a couple quarrelling in the foreground, an Oriental adult when asked, "What is this?", tends to answer "an airport check-in," while an Occidental adult answers "a couple quarrelling," context vs. action.

Oriental thought also emphasizes transience (Buddhism), cycles (Hindu reincarnation), and the unity of opposites (Confucian yin and yang). Pre-determination (karma) and fatalism are the prevalent relations between humans and nature.

The Judeo-Christian-Islamic tradition and monotheism stress eternal truths; promise eternal life. The ideal is separate from the material, and the ideal created the material and determined its development. Humans are created in the image of the ideal. Revelation and reason are part of the ideal and the source of our knowledge. The ideal-human relation, eternal life, and a postmortem Paradise located elsewhere place humans outside/above Nature.

The animists place humans completely at the mercy of deified natural forces who are implored or for whom sacrifices are made. Others consider Mother Earth to be the source of all we need and all we have, our paradise on Earth, with whom we must live in harmony.

The Earth is our heaven, the rest of the Universe is hell for us. This is the worldview most conducive to the struggle for survival and existence. A globalized secular version moves from spirits and divinity to reality with morality, peace among men, and harmony with nature. This position finds spirituality in the evolution of the Universe, planet Earth, and life. It is held by many religious people in the XXIst Century.

Scientific knowledge has logical and methodological rules and rigor, including five explicit epistemological frameworks: empiricism, logical positivism, structuralism, pragmatism, and dialectical materialism. They in turn accept one or another or all of the three modes of explanation: causal-lineal, teleological, and dialectical.

With regard to the question, relevant for ethics, of the role of value in knowledge production, empiricism rejects values that are seen as tainting the objective observation of reality. Logical positivism recognizes the limitations of empiricism's "the facts speak for themselves position" and posits that science should only admit the values of science itself. Pragmatism, structuralism, and dialectical materialism recognize knowledge production as intentional human activity, a practice, and hence a value-laden activity. Wisdom is to be found at the junction of ontology, epistemology, and ethics, combining knowledge and values with ethical or moral frameworks. (See Box 4, "Wisdom.")

Box 4: Wisdom

The ontology of existence and extinction exists independently from human intelligence, but our consciousness is totally within the ontology. Our human consciousness does not condition the transformations of the Universe, but we have and can affect some structures and conditions of Mother Earth, somewhat and to a point. Inasmuch as the expansion of the Universe is structured and conditioned, with a pinch of chance, not random or chaotic, we can know its logic, principles, and evolution.

The nexus between ontology and epistemology is the theory of everything, information. Expansion of the Universe and evolution of Earth are based on information. Our sensory experience is data that interpretation transforms into perceptions that with cognitive interpretation can become knowledge. Our access to the structures and processes of the Universe is based on sensory induction and cognitive deduction in a dialectical unity of opposites.

Our concepts are inexact but can function with "family resemblance," and our logic and methodologies have limitations, but we do produce useful knowledge. The difference between common sense, ideological, and scientific knowledge is quantitative in terms of explicitness and rigor, not a difference in kind. In practice we combine the components, levels of knowledge, and modes of explanation in continuous knowledge production of dynamic, changing reality.

Our knowledge is soçio-historic and value oriented. The combination of knowledge and values organized in an ethical or a moral system is wisdom, what we most lack and need.

The **Ethics of Existence and Extinction** position is that purposive, social activity is intentional and directional, guided by values derived from our historic social and individual objectives. Knowledge production is a purposive, social activity and hence value driven. The evaluation of the social utility of results is value based. Humans e**valu**ate applying values.

It is not that our knowledge is objective and our values subjective but rather that our knowledge and values have the same origin, praxis, the unity of theory and practice.

Values may be **inherent**, contained in the firmware of our genetic inheritance, such as life and survival, or values derived from them, such as safety, wellness, and well-being. Other values are **social**, such as love and solidarity, without which none of us would survive the total helplessness of human infancy and near helplessness of prolonged child rearing. Throughout life we are socially dependent on each other for life and survival.

A third level consists of **constructed** values. Our social practice seeks safety, wellness, and well-being for survival, but we also strive for purpose and fulfillment, constructing needs/values in the process. With one notable exception, we are not a particularly well-endowed species. Our intelligence is our comparative advantage, and it is social intelligence. Learning and knowledge have been and are vital for life and survival, but also for constructed needs and values for achievement, fulfillment, and self-realization by production, creativity, arts, discovery, exploration, collecting, sports, learning, teaching, knowledge, among many others.

Our individual and species's struggle for survival in the life chain of evolution is the **meaning** of our and all life. We should contribute to survival, live life, hopefully to the fullest or at least well, through safety, wellness, and well-being. This is true for all of us.

Self-fulfillment is the **purpose** of life, and this is constructed by each of us as social groups and as individuals according to our needs, interests, capacities, and achievements. Our level of success in these endeavors is judged by others; the level of satisfaction is our own judgement.

A fourth level is **ethical or moral** values. Ethics are value configurations and rules that everyone "should" follow; morals add the imperative of "must" follow backed by social, religious, and/or legal sanctions. Groups have their own objectives, needs, interests, and positions leading to conflicts, exclusion, and domination that can lead to violence and wars. Conflict is inherent in all social life and may be channelled peacefully or violently. Violence is human self-destruction or destruction of others. The values of equality, equity, and justice can reduce exclusion and domination, creating conditions for conflict resolution.

Values, attitudes, beliefs, worldviews, customs, behaviour patterns, and material artefacts, including tools, technologies, utensils, and art, constitute **Culture**. Ethical/moral values, formal/informal rules, including laws, incentive structures, and national/organizational cultures, form **Institutions** that structure our interdependence. Culture and institutions are fundamental in generational/group/individual identities and their evaluations.

Peace was desirable before nuclear weapons; now it is a moral imperative. An accounting of nuclear policy, events spiralling out of control, close calls, accidents, and human failings impress upon us how precarious our existence is with nuclear weapons lying around the planet.

The need for dignity (freedom from humiliation), human rights (freedom from oppression), and human security (freedom from fear) elevate values to the level of ethics/morals. Conspiring against this is the ever more pervasive culture of violence diffused through cinema, television, digital games, and terrorist or official, violence as part of full-spectrum dominance.

Full-spectrum dominance consists of unipolar military and political hegemony, permanent warfare, unilateral declarations of regime illegitimacy, covert or overt actions to destabilize regimes the hegemon doesn't control; regime change wars, negation of international law but extraterritoriality of hegemonic law; unilateral, coercive measures that are mostly illegal (the only legal sanctions are approved by the United Nations Security Council); administrative freezing of assets without judicial recourse; control of financial transfers, negation of banking services; extra-territorial criminalization of third parties that don't comply with the illegal measures; capacity to intercept all communications everywhere, cyberattacks, cyberwarfare, fake news, social media manipulation, media and cultural hegemony, offshore prisons to obviate rights, indefinite detention without trial for foreigners, drone killings by administrative edict, objective of Damocles sword capable of killing anyone on the planet, anywhere, within one hour, as a totalitarian agenda; all for military, political and economic hegemony.

Our generation will be evaluated on our ability to overcome processes now threatening future generations, including permanent warfare, weapons of mass destruction, culture of violence, full-spectrum dominance, climate change, global warming, environmental degradation, and the maximum crime against nature, species extinctions, including our own species's suicide, which continues to take shape.

The knowledge we require for survival is of the philosophy, economics, politics, policies, and practices that lead to survival, based on science, not political compromises, to the lowest-common carbon emission denominator.

The 2018 report of the Intergovernmental Panel on Climate Change (IPCC), the scientific body of the United Nations Climate Change Convention, on how to limit the average annual worldwide temperature to 1.5°C above the pre-industrial level in this century finds the following. First, it is necessary to reduce greenhouse gas emissions by -45% by 2030. Second, we must reach a zero-net emissions, sustainable society by 2050.

If we fail to meet these goals, we will leave a degenerate inter-generational injustice for our children and grandchildren, our hedonistic pleasures and luxuries now for their future of increasing misery and death. Or are we capable of forming a Survival Movement to construct a low-carbon, climate resistant, circular, sustainable society with greater equality? That is the great challenge of this historical epoch that will define our and future generations.

A well-known scientist in a radio interview was asked what he thought was the probability that aliens would visit Earth in our lifetime searching for intelligent life. He answered that his greatest fear was that they had already arrived, searched, and left. It is hard to refute that on environmental degradation, including climate change, as well as nuclear weapons, we have hardly exhibited intelligence as a species, given the threat of self-destruction.

What is needed is the construction of a low-carbon, climate-resistant and resilient, sustainable society. Once again, in our second and probably last chance, Mother Earth can supply all we need but definitely not our greed. We have the financial resources, but they require radical reallocation. The survival movement must mobilize social and political will based on our moral strength and spiritual wherewithal combined with ultimate collective self-interest and altruism for humanity, all species, life itself.

Spiritually we need to understand who we are in the Universe and on planet Earth, an integral part of both, a unique individual of our species within the totality of life and existence. Our greatest opportunity is to participate in life and existence, and our greatest responsibility is to protect life and existence.

The Ultimatum

The Ultimatum

001 The Ultimatum is imperative—transform ourselves now or decline and perish because of our own actions, a species suicide.

002 The Ultimatum is dual: we can produce our own extinction and that of others through weapons of mass destruction, especially nuclear and/or environmental degradation, especially climate change.

003 The Ultimatum is clear from Hiroshima, Nagasaki, Chernobyl, Fukushima, in 4000 operational nuclear warheads, the use and potential use of chemical and biological weapons.

004 The Ultimatum is clear in deforestation, tropical rain forest, and top soil losses, as well as in the advance of arid zones and deserts on all continents.

005 The Ultimatum is clear in the loss of species habitat, especially for large mammals, accelerating loss of biodiversity through mass extinctions.

006 The Ultimatum is clear with climate change, global warming, increased strength and frequency of extreme weather (tropical depressions, storms), tornadoes, flooding, and droughts.

007 The Ultimatum is clear in the melting of polar ice, Greenlandic and mountain glaciers, the rise in sea levels, and the loss of coastal and insular areas.

008 The Ultimatum is clear in declining coral reefs and 90% of large fish stocks lost by overfishing, food chain disruption, ingesting plastics, ocean temperature rise, and acidification.

009 The Ultimatum is clear in reduced surface/ground water and contaminated rivers, lakes and seas, over 150 oxygen-deprived dead zones in the oceans, and islands of plastic refuse.

010 The Ultimatum is clear in ozone depletion, increased ultraviolets; air, water, soil pollution; the poisoning of food, water, other species, land and sea, you and me.

Our Response

011 Our response to nuclear, chemical, and biological weapons, environmental degradation and climate change threats, and loss of EQUILIBRA within/between life support systems is pathetic.

012 Pathetic Ultimatum responses augment risks through business as usual, wrong actions, too little too late, greed, inaction, procrastination, resignation, negation.

013 All species have struggled & continue to struggle for survival because of cosmic, geological, climate, habitat, predator, food, water, & epidemiological threats, our most recent: Novel Coronavirus, H1N1, Ebola, SARS, HIV.

014 What is new in the 20th and 21st centuries is that it is our own past/current/potential actions that are the greatest threats to ourselves and others' existence.

015 **That we ourselves are the threat makes the situation more difficult in that it is easier for us to continue transforming Nature than transform ourselves.**

016 We suffer addiction to unsustainable models of production, consumption, finance, and lifestyles and to reinforcing forms of education and social communications.

017 We have lost touch with our place in the Universe and in Nature on planet Earth, because of alienated anti-values—species centrism, greed.

018 Anti-values are institutionalized in entrenched interests, concentrated power, hegemonic structures of domination that reinforce the anti-values.

019 Despite consciousness of the destructive power of chemical/biological weapons since WWI and of nuclear weapons since WWII, they still have not been abolished.

020 Despite consciousness of crises in bio-diversity, desertification, and climate change since Río 1992, business as usual predominates and the crises advance.

Transformative Capacity

021 Our transformative capacity, manipulative and cognitive, the formula for our species success, now threatens our existence, making change difficult.

022 Our evolutionary journey has consisted of increased physical, cognitive, family, and social capacity to transform the environment for our survival and well-being.

023 The evolution of the genus *homo* to *homo sapiens sapiens* and *homo Neanderthalis* has been one of ever greater cranial capacity/brain and manipulative power.

024 Our bipedal position and opposed thumbs have allowed grasping, holding, carving, throwing, creating, and perfecting tools/weapons, increasing adaptation.

025 A low hyoid bone, tongue, and larynx facilitated language, communications, knowledge formation, storage and retrieval, synthesis/analysis in our species.

026 Prolonged infant child care/juvenile learning enhanced cognitive, information, knowledge, and family capacity for transformative/adaptive actions.

027 Tools, language, knowledge led to social development from the family/extended family to the clan and tribe, increasing social transformative capacity.

028 Increased social capacity, tools, and weaponry moved us beyond gathering, hunting, and fishing to water management, agriculture, food storage.

029 Neolithic settlements achieved food surpluses and division of labor, including farmers, artisans, merchants, priests, military, teachers, and wise men.

030 Civilizations led to commerce, accounting, law, transportation, and eventually industry, science, and technology, increasing social transformative capacity.

Destructive Capacity

031 The same transformative capacity that has been a large part of our success has now reached an inflection point where it becomes a threat to our existence.

032 As we increased our capacity to transform the environment, we also increased our capacity to destroy both the environment and ourselves.

033 Our early transformational capacities caused limited, localized damage to water, soils, forests, habitats, other species with negligible air pollution.

034 Perhaps we overhunted the great mammoths and other species and contributed to the Neanderthals' demise through assimilation, food competition, violence.

035 Through time, forests disappeared in some regions, water was contaminated, soils depleted, and habitats affected by agriculture and urbanization.

036 The increase in our population augmented pressures on the environment in some areas, while our societal responses were isolated, marginal.

037 The rise of mercantilism, European empires, the industrial revolution, capitalism, science, and technology accelerated environmental destructive capacities and a culture of violence.

038 Fossils fueled the industrial revolution and almost continual, unlimited growth of the economies, but they also damage the environment and threaten extinction.

039 Non-renewable resources are being depleted, and renewable resources such as forests, water, soils, flora and fauna are being increasingly degraded.

040 Market greed continues to take precedence over species need in investment in energy, industry, infrastructure, and human settlements, as well as in science and technology.

Impact

041 Endless, limitless, mindless growth, production, consumption, and capital accumulation cannot continue on a planet with limited, degraded resources.

042 Air has lost quality, and we have less fresh water and fewer fertile soils. Food is abundant for the rich and scarce for the poor with high risk of future famines.

043 The economic model produces more refuse than it can handle, including toxic and nuclear waste; 300 tons/year of non-biodegradable plastics of which only 9% have been recycled.

044 Sea habitat and biodiversity are lost by temperature rises, acidification, oxygen depletion, solid/chemical pollution (more plastics than fish by 2050) over fishing, extinctions.

045 Increased cancer and other mass deaths attest to environmental contamination. Bees and frogs are first, but we are also in the line marching to extinction.

046 Cattle grazing, urbanization, urban sprawl, and recreation are leading to habitat and biodiversity losses that have reached the point of mass extinctions.

047 Land and sea extinctions are by areas, in the wild wherein the only surviving individuals are in captivity, and in the form of absolute extinction.

048 We are now living, because of human impact on the environment, in the Sixth Major Extinction event in the history of the planet and the first caused by humans.

049 What are not **yet** visible are negative synergies between and within the life support systems that allowed the emergence of life on planet Earth and now sustain it.

050 Direct damages are visible; damages of weapons of mass destruction are potential and genocidal; inflection points to systems **desequilibra** are dangerously unknown.

Societal Incapacity

051 Life emerged through multiple **EQUILIBRA** within, between ecosystems and positive synergies maintaining life, contributing to evolution, a living planet—Gaia.

052 All we have to do is not interfere with life support **EQUILIBRA**, but we are interfering with geosphere, hydrosphere, atmosphere, and biosphere **EQUILIBRA** and their synergies.

053 That the species success formula of millennia now poses the threat makes recognizing its unsustainability as well as effective responses more difficult.

054 Individuals, nations, our species are slow to act against what has been the source of success upon reaching an inflection point of threats and decline.

055 For 800,000 years prior to 1750, atmospheric greenhouse gases were below 280 ppm; by 1988 they rose to the critical 350 ppm level; in 2013 they passed 400 ppm and in January, 2020, 413 ppm.

056 **The smokestack passed from symbol of success to symbol of global warming/climate change, while the high-energy economy went from indicator of success to sign of excess and waste.**

057 Weaponry and warfare became mass military destruction in World War I and mass civilian destruction in World War II and afterward: permanent warfare and pervasive culture of violence.

058 Hiroshima and Nagasaki demonstrated our capacity to destroy our own species and life on Earth; 4,000 operational, strategic nuclear weapons remain in existence and the risk is real.

059 The hegemonic power structure places a higher priority on financial systems and bailing out banks than ecosystems and saving the synergies that maintain life.

060 **The reason the hegemonic politicians cannot control the monopolists, bankers and speculators is because the monopolists, bankers and speculators control the politicians.**

Our Dilemma

061 Billions of years of structured, conditioned evolution of the Universe and of Mother Earth, as well as miraculous chance, led to our species.

062 The living planet, Mother Earth, emerged and survives in an EQUILIBRA of mutually reinforcing and self-regulating life support ecosystems.

063 **Our conscious species existence and our individual lives on this planet that provide all of our needs is our heaven, the only place in the known Universe for us. The rest is hell.**

064 We are an integral part of a chain of life that depends completely on, as well as reinforces, the self-regulating ecosystems that sustain life.

065 For millennia there was humility, even worship, before the forces of the Universe and of Mother Earth, realizing that we are totally dependent upon them for our existence.

066 Egoism now supersedes traditional humility and rationalizes unsustainable patterns of production and consumption despite consciousness of consequences.

067 Hegemonic culture once held that the Earth was the center of the Universe, but still believes that the Universe revolves around human beings.

068 Species centrism underlies unsustainable patterns of production, consumption, finance, and lifestyles that have our human civilization in accelerating decline.

069 The enormous disruption of lives, societies, economies of the Coronavirus is small, transient and recoverable, compared to the total, permanent, and irreversible damage of a nuclear exchange or the disruption of life support eco-systems.

070 Cosmic and natural evolution that led to us, all the millennia of sacrifice by our ancestors and the future of our progeny are all riding on our actions.

Nine Alienations

071 Hegemonic culture is separated from nature through nine alienations that profoundly influence our psychology and reinforce actions and processes tending toward extinction.

072 The first alienation is to think only the eternal is valuable and escape transient reality by creating eternity in thought, devaluing "now," *eternalism.*

073 The second alienation is to believe that humans are not dependent on Nature, but rather outside of Nature, and even somehow above Nature, *supernaturalism.*

074 The third alienation is to posit humans as the masters of all Nature, which exists to serve us, and degradation as an inevitable cost of progress, *developmentalism.*

075 The fourth alienation is to place individual, group, or national consciousness or interests before our and other species' survival, *ultraindividualism.*

076 A fifth alienation is to prioritize short-term marginal gains above long-term catastrophic losses, even knowing the consequences, *immediatism.*

077 The sixth alienation is to think that endless, limitless, mindless growth and capital accumulation can occur on a planet with limited, degraded resources, *capitalism.*

078 A seventh alienation, part of capitalism, is to prioritize financial systems (commodity, equity, derivative markets) above life support systems, *financialism.*

079 An eighth alienation is to put all faith in science and technology to replace or compensate for damages we cause to ecosystems/environmental **desequilibra**, *scientificism.*

080 The ninth alienation is *fatalism* with two variants, providentialism and nihilism, one resigned and passive and the other negative, obstructive and prone to violence, respectively.

Survival

081 The ecosystems that led to, and support, life on Mother Earth are adaptive, resilient, synergistic, self-regulating, and reproducing, a living system themselves.

082 Positive synergies reinforce the existence of life on Earth, but there are unknown structural inflection points that will turn them negative.

083 Life on Earth is based on a set of **EQUILIBRA** of life support ecosystems, as is also the case for the existence of a given species or of its individuals.

084 Defense of **EQUILIBRA** requires species consciousness, action, and the transformation of political, economic, and behavioral structures that weaken existence.

085 Political and economic structural transformation requires overcoming cultural, immoral, anti-ethical, and anti-value structures; transform our species to save our species.

086 To transform ourselves we must construct a species consciousness based on knowledge of our history and our place in the Multiverse, Universe, life on Earth.

087 The knowledge we require is of the philosophy, economics, politics, policies, and practices based on science, not political compromises to the lowest common carbon-reduction denominator.

088 Science: Halve CO_2(e) emissions every decade from 2020 to 2050 for a 66% chance to limit the temperature increase to 2°C and 50% for 1.5°C; otherwise, it's a catastrophic 3-6°C.

089 The IPCC 2018 report limiting the world temperature rise to 1.5°C indicates we must reduce emissions by -45% by 2030 and achieve a zero net emissions, sustainable society by 2050.

090 We must construct a Survival Social Movement for a low-carbon, climate-resistant, recyclable, sustainable society with far greater equality. This will define our generation.

Social Movements

091 Social transformations are achieved by social movements: anti-slavery, anti-colonialism, racial equality, labor, gender, peace, green, and now survival.

092 Social movements are networks of people in different countries, occupations, sectors, working for the same cause with similar objectives and historical perseverance.

093 Social movements are based on a common consciousness of an injustice and/or threat and the formation of a community to spread that consciousness to others.

094 The injustice or threat is always a moral and ethical issue because it adversely affects millions of humans or all of humanity or all of life on Earth, an existential level.

095 The moral and ethical dimensions of movements are based on values such as life, love, survival, equality, equity, justice, among others, which imbue them with spiritualism.

096 The value, ethical, and moral dimensions of social movements make them similar to religious movements that almost always participate in them.

097 Social movements consolidate when they articulate their consciousness, encounter effective forms of action, and aggregate inclusive support.

098 Social movements weaken when they encapsulate into sects that exclude rather than include and cannot communicate their messages effectively.

099 Social movements succeed when their cause is incorporated into social policies, including laws, and eventually into universal species social consciousness.

100 A social movement for survival to overcome the twin nuclear and environmental threats, our alienations, and the deformed power structure is our best, last hope.

Evolutions/Revolutions

101 While social movements are value, ethically, morally, and cause driven, political parties seek to capture and exercise power; movements need to influence parties and policy.

102 Some political parties embody the cause of a movement, on occasion multiple parties adopt the cause, and with consolidation all parties support.

103 The interactions and dynamics between movements, parties, states, and international organizations take place within power structures.

104 Power structures are based on correlations of forces that are transformed by structured, conditioned transformations and changing conditions.

105 When a social movement achieves its social policy objectives within the framework of a given power structure, it is a social evolution.

106 When the achievement of the social policy objectives takes place only with the transformation of the power structure, it is a social revolution.

107 Our dilemma resides in unsustainable patterns of production/consumption/finance/lifestyles that are the foundation of the hegemonic power structure.

108 The hegemonic power structure is constructing a military/industrial/intelligence/police complex with global reach that aspires to full-spectrum dominance.

109 Full-spectrum dominance aspires to totalitarian control: unipolar military, political, financial, media, communications and cultural hegemony, permanent warfare.

110 We must transform ourselves, our species's consciousness, and a repressive hegemonic power structure to advance and consolidate our Movement for Survival.

The Philosophy of Existence and Extinction

The Ultimatum and the Philosophy of Existence and Extinction

Metaphysics, Ontology, Epistemology, Ethics

Logic of the Multiverse

111 The Logic of the Multiverse is structured, not random, consisting of lifecycles of structured, conditioned transformations, with a pinch of chance.

112 Lifecycles have a beginning, a process, and an end that characterize everything in the Multiverse with only one possible exception, the Multiverse itself.

113 The succession of multiple universes within the Multiverse is the only possible exception to the change constant, far beyond our comprehension.

114 The outcomes of structured transformations within their possibilities and limits are determined by interactions with entities/processes known as conditions.

115 Structural transformational possibilities and the conditions that determine processes, structures, as well as chance outcomes, are sets of information.

116 Chance events occur when structured outcomes are conditioned by competing sets of conditions, indeterminacy, especially in changing conditions.

117 Chance can be temporal ("when")/existential ("if"). Temporal may condition when you may die, not if. In if, Earth's chance distance to Sun permits life conditions.

118 Structured, conditioned transformations evolve from simplicity to complexity, homogeneity to heterogeneity, including the Universe, life, you, and me.

119 There is no exception to the logic of the Multiverse that existence/extinction and life/death form a unity of opposites; each implies the other.

120 **Structured transformations within lifecycles determine** the evolution of the Universe, life on Earth, our species, your life, my life.

The Principles of the Multiverse

121 The logic contains principles inherent in the conditioned, structured transformations of the evolution of all entities within their lifecycles.

122 The core unity of opposites is **EQUILIBRA/desequilibra**, which structures forces, energy, and matter in the Universe and their interactions.

123 The Universe began with the desequilibria of the Great Expansion of the Singularity in space-time that is our structured, conditioned Universe.

124 Existence within the Universe depends on an **EQUILIBRA** of forces, energy, and matter in processes, entities, structures, and systems.

125 Life on Earth began with an **EQUILIBRA** of the forces, energy, and matter of the cosmos, litho, magneto, atmosphere and hydrosphere creating the biosphere.

126 The **EQUILIBRA** of the evolution of the biosphere has been facilitated by self-regulating synergies between all the spheres forming ecosystems.

127 The self-regulating **EQUILIBRA** is homeostatic within the parameters of existence actively maintaining a living system, Gaia, Mother Earth.

128 The evolution of life led to our species *homo sapiens sapiens* within the living **EQUILIBRA** of Mother Earth as an integral part of the Universe.

129 The logic and principles construct and deconstruct entities and processes through forces, energy, and matter in time within their lifecycles as per information.

130 **EQUILIBRA/desequilibra** is the foundation of the principles of The Unity of Opposites, the Change Constant, Identity in Diversity, and Information.

The Unity of Opposites

131 Forces, energy, and mass are all structured by the Principle of the Unity of Opposites, as is their integration and disintegration.

132 The integration and disintegration of energy as well as energy/mass is structured by the Unity of Opposites of EQUILIBRIA/disequilibria.

133 The **desequilibra** of the implosion/explosion of the Singularity and conditions of different densities/velocities/temperatures unfolded energy/forces/matter.

134 The initial expansion was at a density, velocity, and temperature that was pure energy, light energy and dark energy, a structuring unity of opposites.

135 The first force to appear was gravity, whose opposite is weightlessness, and the second was the strong force, whose opposite unity is the weak force.

136 The electromagnetic force unfolded, consisting of electromagnetic attractions and electromagnetic repulsion, positive and negative poles.

137 Energy particles without mass include photons/anti-photons, quarks/anti-quarks, neutrinos/anti-neutrinos, whose formation/liquidation structured early Expansion.

138 There is matter and dark matter. Particles with mass include positrons with positive charge and their opposite, the electron with negative charge.

139 Matter-energy and energy-matter conversions conditioned by forces form a fundamental unity of opposites beginning from the origin of our Universe.

140 Existence and extinction are a structuring unity of opposites for life on Earth and species life, while life and death are our individual equivalents.

Change and Simplicity to Complexity Constants

141 Change is a constant in that everything changes. The only possible exception is the Multiverse itself and far beyond our comprehension. Life began and begins with a single cell.

142 Given the unity of opposites and the change constant, everything has a beginning and an end, including you and me. All evolves from simplicity to complexity. All is transient.

143 You and I have had a beginning, birth, and will have an end, death. Our species came into existence, evolved, and will become extinct.

144 Life on Earth came into existence, has evolved, undergone major extinction events, moving from the simple to the complex, homogeneity to heterogeneity, and will become extinct.

145 The Sun will run out of fuel and stop converting hydrogen into helium by fusion, expand to be a red giant, then collapse as a compact white dwarf in some 5.4 billion years.

146 Our galaxy, the Milky Way, will merge with Andromeda and eventually disappear into the gravitational pull of the combined black holes.

147 Galaxies, nebulae, forces, energy, and matter, including particles, had a beginning and will have an end. Particles/anti-particles cancel out.

148 The Universe began with a primordial Singularity, maximum simplicity, and could end by the implosion of black hole singularities into a new primordial Singularity.

149 The universe could expand to the point of losing systemic structure and existence. A corollary is expansion into a dead absolute zero corpse.

150 The end depends on if accelerating dark-energy expansion remains dominant or if dark-matter black holes coalesce and collapse all, but all will end.

Identity in Diversity

151 All entities have a unique essence/identity in the diversity of their types. There are no identical galaxies, stars, planets, humans (fingerprints), snowflakes, sand.

152 Combinations of entities include structures and systems of entities. They all have a defining essence that is the common identity in diversity.

153 All individual entities may be grouped into types. They all share the identity of essence of a type and a distinct identity as individuals.

154 There are different types and individuals of planets, stars, stellar systems, galaxies, and we can deduce even universes within the Multiverse.

155 No two individuals of any of the types of the Multiverse are exactly the same. Every individual is a unique and co-equal part of the Multiverse.

156 A segment of space time in EQUILIBRA is an entity, type of entity, or individual. This is the case for galaxies, stars, planets, life, species, individuals.

157 All entities, types, and individuals exist in an EQUILIBRA in the space-time of a given universe of the Multiverse for the time of its lifecycles, as per information.

158 A type or individual becomes extinct or dies with **desequilibra** at a point in space-time of a specific Universe of the Multiverse as per the information of its lifecycle.

159 There is nothing more important than maintaining the EQUILIBRA of ecosystems that support our species's existence and our individual lives in space-time.

160 Our end will come, and a lapse of space-time is all we have. Human impact on the Earth is now shortening our space-time, threatening the loss of all we have.

Information as the Theory of Everything

161 Information is contained in and structures the Universe, including the principles of unity of opposites, change and simplicity to complexity constants, and identity in diversity.

162 The "cosmic stem cell" of the Singularity contained all of the logic, principles, energy, forces, matter, and energy of the Universe as information.

163 "Cosmic DNA information" structured evolution through the definition of potentialities of entities and their evolution or not, as per information on conditions.

164 At reducing temperatures/speeds/densities, energy, forces, different forms of matter decoupled and later combined as per information.

165 Different types of stars are constructed and destructed according to the information of the lifecycle of that type of star, influenced by others through the information of gravitational attractions.

166 When a star is swallowed by the gravitational attraction of a black-hole singularity it leaves behind information of its previous existence.

167 Organic molecules read other organic molecules and react accordingly. Our immune system reads the markers of bacterial and viral pathogens and reacts accordingly.

168 Genetic inheritance information is stored in DNA and communicated by RNA and proteins to cells in the process of formation of living organisms.

169 Human beings are conceived when the information of 23 male chromosomes is combined with the information of 23 female chromosomes. A single cell divides & begins to communicate.

170 **Life reads, is structured by, and adapts to information, as well as maintains homeostatic EQUILIBRA by processing information that is the theory of everything.**

ONTOLOGY OF EXISTENCE AND EXTINCTION

Cosmology, Existence, and Extinction

Cosmology

171 The Singularity was a dimensionless, zero-volume, infinitely dense point at absolute zero in maximum curvature of space-time for the most minimal moment.

172 The Singularity was formed by the implosion of the black holes of prior Universe(s) in the process of construction/destruction of the Multiverse.

173 The Singularity was not ultimate stability but rather ultimate instability in that upon imploding it expanded. Change is the only constant.

174 The Singularity was the stem cell that contained the "cosmic DNA" of the structured, conditioned transformations of the lifecycle of the Universe.

175 The Singularity contained the information of all the energy/forces/matter and of the conditions of their structural transformations within the logic/principles.

176 The unity of opposites, change constant, unity in diversity structure reality based on information on types/individuals/transformational possibilities.

177 Force information initiates processes and structural transformations within possibilities, but condition information determines specific outcomes.

178 The primordial condition was velocity of expansion/energy/ temperature, and the first processes and structures were conditioned by slowing/cooling.

179 Temperature also conditions stellar fusion and the creation of heavy elements, as well as the **EQUILIBRA** of the ecosystems that sustain life/us.

180 We have been ill advised to largely ignore the 1988 rise in greenhouse gases above 350 ppm, the 2013 400 ppm levels, and the decade of 2010 as the hottest years on record.

EXISTENCE:

Great Expansion, Photons, Cosmic Microwave Background

181 Existence began with the implosion/explosion (Great Expansion) of the Singularity some 13.77 billion years ago; that is the Universe we inhabit.

182 At the maximum temperature the Universe has ever known, 100 billion degrees Kelvin, there was a Single Unified Force and no matter, only pure energy.

183 Pure energy was carried by the photon, at once a light/energy wave and a weightless elementary particle without charge that is its own opposite.

184 Photons travel at the speed of light, and hence space and time are measured relative to the photon. The photon is the basis of all our metrics.

185 The photon is the quantum of light and the carrier of electromagnetic radiation as well as the gauge boson that carries the electromagnetic force.

186 Cosmic Microwave Background (CMB) from 380,000 years into the Expansion exists throughout the Universe, cooled to +2.725 above absolute zero.

187 CMB absorbs 99.9% of all protons with an intensity of 300 photons per cm3 of the Universe. Only 0.1% of light/radiation is in galaxies/stars.

188 CMB is the oldest, most prevalent, coldest thing/wave in the Universe, evidence of our origin 13.7 bya, and of subsequent transformations of the early Universe.

189 With rapid Expansion volume increased and density, velocity, and temperature decreased, reaching conditions for structured transformations.

190 First the forces unfolded accompanied by the gauge bosons that mediate their actions and interactions, then other bosons and finally elementary fermions.

EXISTENCE:

Forces, Bosons

191 At the right mass/energy/velocity/temperature, gravity unfolded followed, when the information was right, by the strong force that integrates atoms.

192 There followed the weak force that decays and disintegrates atoms and the electromagnetic force with its positive and negative charges.

193 Photons are the gauge boson for the electromagnetic force. The graviton boson is a massless wave that mediates gravity, as of yet unobserved.

194 Pair production of photons at different densities, velocities, temperatures produced the other gauge bosons that intermediate the forces.

195 At the strong force symmetry breaking threshold, photon collisions produce the 8 gluon bosons with different masses that mediate atom binding.

196 At the weak force symmetry breaking threshold, photon collisions produce the weak force mediation bosons W+, W-, Z, which decay/disintegrate atoms.

197 Similarities between the Electromagnetic and Weak Forces have led to the hypothesis of an underlying Macro Weak Electromagnetic Force.

198 However, there is a difference in the gauge bosons, the photon has no mass, but the bosons W and Z have 100 times the mass of a proton.

199 Interaction in the Higgs Field and with the Higgs Boson are posited as creating mass in W and Z bosons, whereas Photons do not have this relation.

200 The full presence of the force fields and the gauge bosons facilitated the acceleration of the formation of fermions, hadrons, leptons.

EXISTENCE:

Fermions, Hadrons, Leptons

201 Photon collisions can also produce the fermion lepton pairs electrons and positrons, as well as fermion hadron protons and anti-protons.

202 Fermions consist of invisible quarks joined in hadrons, the basis of baryonic matter, including protons and neutrons, linked to the strong force.

203 Fermions also consist of leptons such as electrons and neutrinos, components of the atom that are linked to the weak force.

204 At one pico second the Universe consisted of a quark-gluon plasma of elementary particle and anti-particle pairs created and destroyed in collisions.

205 At one microsecond quarks combined with hadrons whose epoch lasted until at one second, and 10 billion-degree leptons became dominant.

206 Chance excess of quark/leptons over opposites led to quark/gluons forming hadrons/baryons (protons/neutrons) whose baryongenesis had chance excess of matter of 1 in 10 billion.

207 From 10 seconds through 379,000 years, the Photon Epoch produced the sub-atomic building blocks for nuclei, accelerated by dark matter.

208 One hundred seconds into the Expansion at 1 billion degrees Kelvin, protons and neutrons did not have enough energy to resist the Strong Force.

209 Collisions produced particle pairs, including the opposite electrons/positrons that have mass derived from the energy of formative collisions.

210 Our species, you and me, have our origins in the Great Expansion, which generated the energy, forces, and particles that formed the elements of the universe.

EXISTENCE:

Elements, Atoms, Mass

211 In strong force fields protons and neutrons combined to form the nuclei of deuterium, helium, hydrogen, lithium, and beryllium.

212 In the process of nucleosynthesis the Strong Force bound pairs of one neutron and one proton to form deuterium nuclei (heavy hydrogen).

213 Deuterium nuclei combined with other neutrons and protons to form helium nuclei formed by two neutrons and two protons, then lithium and beryllium.

214 Remaining neutrons decayed into protons and most formed the one neutron nuclei of ordinary hydrogen, which remains the most prevalent element in the universe.

215 Baryonic matter/energy of which we are part composes only 4.6% of the Universe, while cold dark matter is 23%, dark energy 73%, and neutrinos 0.4%.

216 At 379,000 years and at a few thousand degrees, nuclei no longer had the energy to overcome the electromagnetic attraction between them.

217 Electrons and nuclei combined into atoms, mostly hydrogen and less helium, which still predominate to date, especially in stellar fusion.

218 With cooling the mass energy of baryonic matter gravitationally dominated that of photon radiation, ending the Photon Epoch.

219 Photon radiation without mass decoupled from matter and moved unimpeded through space with an even distribution throughout the Universe.

220 The building blocks of the Universe are the same as those of life on Earth, and both lifecycles started small and simple and then evolved.

EXISTENCE:

Asymmetries, Galaxies, Stars, Nucleosynthesis

221 The asymmetry of baryonic particles over their anti-particles led to some matter surviving rather than being completely annihilated by anti-matter.

222 The Universe began as homogeneous and isotropic and largely remains so, but with cooling, matter in some areas was denser than the average.

223 In denser areas the Expansion was slowed by stronger gravitational attraction and in some areas stopped, causing the collapse of denser matter.

224 Collapsing matter formed galaxies and external gravitational attractions from quasars started some to spin, forming rotating, spiral galaxies.

225 Collapsing matter not attracted to rotation became oval, elliptical galaxies with rotation around the center but no overall rotation.

226 Within galaxies, clouds of hydrogen/helium break into smaller clouds, imploding because of their own gravity, while collisions raise temperatures.

227 Temperature increases in imploding gas clouds generate nuclear fusion of hydrogen into deuterium and helium producing heat, light, stars.

228 After hydrogen fusion, stellar cores contract, energy/temperature levels rise, and helium fusion nucleosynthesis creates heavier elements.

229 Helium fusion creates carbon, which combines with helium to create oxygen, which combines with carbon to create neon, then magnesium, silicon, and iron.

230 Iron that is highly stable and not subject to fusion accumulates at the core of large-mass stars until distribution by supernova explosions.

EXISTENCE:

Stellar Wind, Supernovas, Black Holes, Stellar Cycle

231 In low-mass stars nuclear synthesis reduces the core and increases energy and temperature levels, causing uncontrolled helium fusion hot flashes.

232 Heavy elements rise to the surface of low-mass stars by convection during helium flashes, while strong solar winds feed the local Universe.

233 The collapse of the iron core of large-mass stars creates supernova explosions that produce more energy and light than an entire galaxy.

234 Supernova explosions project neutrinos and electromagnetic pulses throughout the Universe as well as heavy elements into nearby nebulae.

235 Supernova explosions leave black-hole singularities whose gravity captures other stars, even light, leaving only information.

236 At the center of every galaxy there is a black-hole singularity that grows in mass with the galaxy and whose strong gravity attracts matter and gases.

237 Low-mass star solar winds, high-mass supernova explosions, and black hole gravity all contribute to the formation of new stars/planets in the stellar cycle.

238 The Expansion provided hydrogen (74% of baryonic matter) and helium (24%), while nucleosynthesis provides the heavier elements (2%).

239 The most common heavier elements by mass are oxygen, carbon, neon, iron, nitrogen, silicon, magnesium, and sulfur, which are highly reactive.

240 Nebulae reactions include water vapor, carbon monoxide, and organic matter: formaldehyde, methanol, dimethyl ether, hydrogen cyanide, among others.

EXISTENCE:

Universe, Milky Way, Solar System

241 The expansion of the observable Universe now has a diameter of 93 billion light years with 7,350 billion galaxies and 300 trillion stars (23 zeros).

242 There are 10 million superclusters, 25 billion groups, 350 billion large and 7 trillion dwarf galaxies, including numerous red dwarfs.

243 The supercluster Virgo (100 galaxy groups) contains the "local" group (54 galaxies), including our galaxy the Milky Way, Andromeda, and the smaller Triangulum.

244 The Milky Way and Andromeda are barred spirals with arms in a binary system that will collide in 3.75 billion years, forming a supergiant elliptical with 2 black holes.

245 Intense heat created small planetesimals of rare metals with high melting points (0.6% of the nebulae: iron, nickel, aluminum, rocky silicates).

246 In a 100,000-year evolution through continual collisions, the rocky, metallic terrestrial planets were formed: Mercury, Venus, Earth, Mars.

247 The Jovian gas giants lie beyond the frost line that allows ices to remain solid, capturing hydrogen and helium and achieving large size.

248 The gas giants formed and shifted in a 3-million-year evolution, capturing 99% of sun-orbiting mass and forming Jupiter, Saturn, Uranus, and Neptune.

249 The combined gravitational attractions of Jupiter and Venus make Earth's orbit temporarily 5% elliptical every 405,000 years with environmental impacts.

250 The formation of the planet Earth at the appropriate distance from the Sun for life was a chance factor conditioning the possibility of our existence.

EXISTENCE:

Earth, Moon, Magnetosphere, Lithosphere

251 Our planet Earth was formed 4.54 billion years ago (bya), in an elliptical orbit of the Sun at a distance of 147,093,602 km at the perihelion and 152,097,053 km. at the Aphelion (+3.3%) with an axis tilted 23.5° that causes the seasons.

252 The early Earth was molten iron/nickel whose core solidified by gravity creating a magnetosphere halfway to the Moon, protecting the Earth from some solar wind.

253 The surface of the molten mass around the solid center cooled to form the first lithosphere subject to constant cosmic bombardment and high volcanism.

254 A glancing collision with the large protoplanet Theia vaporized the mantle, forming a molten collision disk from which the Moon was formed by accretion.

255 The Moon has a small iron core, post-molten material, and rocks with the same stable isotope ratio of Earth rocks, evidencing common origin.

256 For 600 million years, Earth and Moon were subject to heavy, continual cosmic bombardment by rays/comets/asteroids/meteorites, adding elements.

257 This bombardment continues to date at a much-reduced rate, yet it is always a threat. We are partially protected by the ozone layer/atmosphere.

258 Volcanism continues to expel heavy elements and gases into the atmosphere and onto the lithosphere at a reduced rate, yet it is always a threat.

259 As a result of continued cooling a viscous asthenasphere formed, and on top a solid surface divided into tectonic plates, a second lithosphere.

260 Movements of tectonic plates have created supercontinents and faults, plate subduction, volcanic vents, earthquake and tsunamis risks and events.

EXISTENCE:

Hydrosphere

261 Hydrogen (H) and oxygen (O) are the #1 and #3 elements in the Universe that combine to form water (H20) at speeds generating 200 degrees Fahrenheit.

262 Water (H2O) exists in vapor/solid/liquid form throughout the Universe. An area of Orion nebulae creates water equal to our oceans every 24 minutes.

263 The Orion protoplanetary disk from which the solar system was formed impregnated water in cosmic dust; 40% of comets and about 22% of some meteorites are H20.

264 The asteroid belt between Mars and Jupiter is principally ice, as are the rings of Jupiter and Neptune. The entire solar system is rich in water.

265 High-energy accretion accumulated volatiles, including H and O, that were degassed as water vapor from volcanoes, as well as delivery by comets, asteroids, and meteorites.

266 Earth cooling led to condensation and the early formation of weather and the oceans/sea lithosphere that now covers 70.8% of Earth with 97.5% of the water.

267 Ocean water evaporates because of the Sun's energy, rains back to the sea or land, and returns to the sea by gravity in the hydrographic cycle.

268 Land water is 2.5% fresh: 68.7% in polar and glacial ice, 30.1% ground water, 0.3% surface water, an atmospheric trace; less than 1% of water is potable.

269 The oldest rocks on Earth are from 4 gya but already rounded by water, demonstrating that the hydrosphere existed since earliest (Hadean) times.

270 The hydrosphere was conceived in Orion, grew out of volcanic lithosphere/meteors, and entered the atmosphere from which it condensed into liquid.

EXISTENCE:

Atmosphere

271 The Earth has had three atmospheres. The first was composed of hydrogen and helium like most in the universe, largely blown away by solar wind.

272 The second atmosphere was formed by volcanic emissions and asteroids 3.4 billion years ago and consisted of nitrogen, CO_2, H_2O, and inert gases.

273 CO_2 was reduced in water, forming silicates, and on land by cyanobacteria using CO_2, H_20, and sunshine for photosynthesis to produce carbohydrates and O_2.

274 Early oxygen combined with lithosphere iron to form iron oxide soils and in water to form iron seafloor bands, constituting 20% of the lithosphere.

275 After iron oxidation was mostly completed, O_2 accumulated in the atmosphere reaching 30%, then reducing to 20%, creating the third atmosphere.

276 Oxygen O_2 was split by ultraviolet rays and single O molecules combined with O_2 molecules to form Ozone (O_3) in a protective layer in the ionosphere.

277 The third atmosphere today is 78.09% nitrogen, 20.95% oxygen (much more than universe), 0.93% argon, trace gases, and H_2O vapor at lower altitudes.

278 Traces are greenhouse gases: $CO2$, methane, nitrous oxide, ozone; neon/iodine/krypton/ammonia/xenon; hydrogen/helium (much less than universe).

279 The atmosphere consists of: troposphere (12 kilometers km high); stratosphere (50km); mesosphere (80km) and thermosphere (320km), together ionosphere.

280 The atmosphere protects partially from meteors, ionized solar energy, ultraviolet rays, retains heat, and contains O2, CO2, and H20, making life possible.

EXISTENCE:

Biosphere, Six Elements of Life

281 The biosphere stems from the Singularity, Great Expansion, stellar fusion, supernova explosions, the Orion proto-planetary disk, the Sun, the Earth.

282 The biosphere as we know it is possible because of the prior existence of the magneto, litho, hydro, and atmospheres and their past and present synergies.

283 The biosphere is based on six elements, five common in the Universe—hydrogen, oxygen, carbon, nitrogen, sulfur—and one rarer element, phosphorous.

284 Hydrogen and helium were formed in the first minutes of the super hot Expansion along with traces of deuterium (H2), helium 3, and lithium 7.

285 Some 200 billion years later, stellar proton-proton fusion reactions began burning hydrogen into helium; fuel exhaustion leads to helium fusion.

286 Helium fuses a helium 4 nuclei (alpha) into beryllium 8 and another helium 4 to create a stable Carbon 12 in the Triple-Alpha process.

287 Hydrogen to helium fusion produces energy and heavier elements, including nitrogen. Helium fusion and carbon production produce oxygen.

288 Sulfur and phosphorous are products of supernova explosions as are the trace elements silicon, chlorine, argon, sodium, calcium, and heavy metals.

289 Cells are formed of 62% oxygen, 20% carbon, 10% nitrogen, 3% hydrogen, 1.14% phosphorous, .16% chlorine, .14% sulfur, .11% potassium, .10% sodium.

290 The progression from lighter to heavier elements and simple combinations to greater complexity constructed a carbon-based biosphere based on the principles of the Universe.

EXISTENCE:

Carbons

291 Hydrogen, oxygen, and carbon are highly volatile. Hydrogen and oxygen combine to form water (H20) and carbon and oxygen, carbon monoxide, and carbon dioxide.

292 Carbon monoxide, carbon dioxide, and diamonds are carbon-based inorganic molecules. All combinations of carbon with other elements are organic molecules.

293 The combination of hydrogen and carbon creates different hydrocarbons of varying complexity, the simplest with 1 to 10 carbon atoms and 1 to 3 bonds.

294 Methane is the simplest of all hydrocarbons with 1 carbon atom/1 bond, ethane has 2 atoms/1 bond; both are found in Neptune's moon Titan.

295 Complex PAHs (Polycyclic Aromatic Hydrocarbons) and even 60 carbon Buckminsterfullerenes (C60) formed by 20 hexagons and 16 pentagons are in the Interstellar Medium.

296 Polycyclic Aromatic Hydrocarbons (PAHs) were formed early in the Great Expansion and are some 20% of total carbon, concentrated in nebulae.

297 In Interstellar Medium conditions, PAHs are subject to and transformed by hydrogenation, oxygenation, and hydroxylation, forming even more complex hydrocarbons.

298 Organic molecules in Universe spectography, comets/ meteorites, include carboxyl and amino acids, including glycine, uracil, glycolaldehyde sugar. Our basic elements are found throughout the universe.

299 Uracil is one of the four nitrogenous nucleobases in RNA, and glycolaldehyde sugar reacts with propenal to form the ribose component of RNA.

300 **The building blocks of life were formed in stellar dust disk, and by accretion, delivered by comets, asteroids, and meteors, as well as generated on Earth.**

EXISTENCE:

Life

301 Life is a microcosm of the Great Expansion of the Universe in the Multiverse at the biosphere, species, and individual levels within the logic of the Universe as per information.

302 Life is formed of the integration of energy, matter, and information in time and space in structured, conditioned transformations from simple to complex.

303 Life is based on a structure that integrates energy, self-replication, and evolution in internal and external **EQUILIBRA**. Death is **disequilibria** and disintegration.

304 Conditions impact life directly and indirectly in structured transformations. Structures and processes act, react, interact, and adapt.

305 Each entity or species develops within a defined lifecycle. This is basic structure. A seed is structured to become a plant or tree in the right conditions of soil, water, sunlight.

306 All of the elements needed for life are to be found in the interstellar medium. This includes the components of DNA and RNA. They are even structured in similar fashion.

307 The self-ordering stacks of **Polycyclic aromatic hydrocarbons (PAHs)** have a separation of 0.34 nanometers, the same separation as between DNA y RNA nucleotides.

308 Homeostatic equilibria proactively seek to keep key survival systemic variables within acceptable parameters to avoid acute system or chronic parameter failures.

309 The elements and macro-molecules of life, matter and structure, are integrated with energy and information in Equilibria and reinforced by synergies.

310 The Great Expansion began with the implosion and expansion of the Singularity. Life began with the Great Combination of energy, matter, structure, and information in RNA and DNA.

EXISTENCE:

Four Macromolecules of Life

311 Varied combinations of the six elements of life and others produce the four macromolecules of life: carbohydrates, proteins, lipids, nucleic acids.

312 Glucose, fructose monosaccharides; maltose disaccharides; starch, glycogen, and cellulose polysaccharides are the carbohydrates of life. All living cells contain them.

313 Glucose, fructose, and maltose provide energy to cells, while starch, glycogen, and cellulose are energy storage molecules. Proteins and lipids aid structures and processes.

314 Amino acids are the most basic units of a polypeptide with carbon, carboxyl, amino, hydrogen, and R-group atoms at four levels, primary to quaternary.

315 Amino acids are the most basic monomer of a protein and the most abundant of the organic molecules, representing 50% of a cell's dry weight.

316 Proteins are directly coded by DNA and provide a cell-structural support, storage, transport, signaling, movement, basic requisites of life.

317 Proteins further provide cellular response to chemical stimuli, catalysis of biochemical reactions, and defense against pathological organisms.

318 Lipid molecules include fats, waxes, sterols, fat-soluble vitamins (A, D, E, K) monoglycerides, diglycerides, triglycerides, phospholipids, others.

319 Lipids provide structural components for cell membranes, energy storage, signaling, transport through membranes, lipid biosynthesis/biodegradation.

320 Nucleic acids are biopolymers that exist in all living cells and function in encoding, transmitting, and expressing genetic information. They constitute bio-infrastructure.

EXISTENCE:

PAH, Nucleobases, Nucleosides, Nucleotides

321 **PAH carbons have formaldehyde backbones with self-ordering stacks with 0.34 nanometer (nm) separation, the same separation as between RNA and DNA nucleotides.**

322 Smaller modules attach naturally to adjacent rings of self-ordering stacks that may swivel in formation, leaving only flat molecules in place.

323 Pyrimidine and purine nucleobases are flat molecules that line up in paired stacks to carry genetic information and perform enzymatic functions.

324 Pyrimidines and purines are nitrogen-based heterocyclic aromatic carbons. Pyrimidine rings are six sided; purine fuses pyrimidine/imidazole rings.

325 Nucleic acids are polymers that are constructed from monomer nucleobases, nucleosides, and nucleotides in strands that are single/double/double helix.

326 Nucleobases are nitrogenous bases that when linked to a pentose (5-carbon) sugar form nucleosides and with 1 to 3 phosphates added become nucleotides.

327 There are two groups of nucleotides classified by their skeletons: purines (adenine, A, guanine G) and pyrimidines (cytosine C, uracil U, thymine T).

328 Nucleic acids pair a purine with a pyrimidine and stack them in strands. RNA has A-U and G-C combinations; DNA has A-T and G-C pairing.

329 The backbone of RNA is a 5-carbon ribose sugar with phosphate, while the backbone of DNA is a 5-carbon deoxyribonucleic acid (DNA) with phosphate.

330 Sequenced nucleotide-stacked pairs encode, store, replicate, transmit, and inscribe information by and for reproduction and protein synthesis/enzymatic functions.

EXISTENCE:

Genome, Chromosomes, Genes, Meiosis

331 The genome is genetic code information written in a sequence of base paired nucleotides A-U for G-C for RNA and A-T and G-C for DNA.

332 The number of base pairs vary widely from life form to life form and species to species as do the number of chromosomes, which are chromatin, DNA-protein packages.

333 Chromosomes consist of genes, the molecular unit of heredity with regard to a specific polypeptide for cell maintenance, genetic traits, and biochemical processes.

334 Chromosomes consist of autosomes for non-sexual reproduction and traits and allosomes, which determine sex in sexual reproduction.

335 A gene is a locatable region of genomic sequence that is a unit of inheritance and an allele is a variant of a trait or process. Genetic sequences contain alleles.

336 Genes determine the genotype of a species, while alleles define the phenotype of an individual within that species based on heredity and recombination.

337 In human reproduction each parent has 23 chromosomes, 22 autosomes for other traits than sex with the 23rd allosome for determining sex.

338 Through the process meiosis, four genetically unique cells are produced with half the number of chromosomes as in the parent.

339 Meiosis also produces combinations each with half the number of chromosomes of each parent to produce a gamete with a new phenotype for diversity/adaptation.

340 The halved chromosomes allow gametes (sperm and egg) to fuse with fertilization to form a zygote with a mix of 23 paternal and 23 maternal chromosome combos.

EXISTENCE:

Mitosis and Genetic Expression

341 In mitosis chromosomes divide into two daughters that combine for asexual reproduction/cell division and growth/replacement/regeneration in plants and animals.

342 Gene expression also contributes to morphogenesis, cellular control of structure and function, cell differentiation, versatility/adaptability/evolution.

343 Gene expression synthesizes utilizable genetic products in the form of proteins; transfer, small nuclear, and messenger RNA (tRNA, snRNA, mRNA).

344 Gene expression processes are modulated, including transcription, translation, post-translation gene protein modification, and RNA splicing.

345 Transcription consists of the production of RNA copies of DNA by RNA polymerase, which adds one RNA nucleotide at a time to an expanding strand.

346 Translation converts the meaning of a source language into a target language, as when RNA instructs ribosome amino acid to protein conversion.

347 Posttranslational modification is part of protein biosynthesis, which can incorporate 20 amino acids that can be folded, cut, and spliced to finish the product.

348 The ribosome translates an mRNA sequence into insulin, which is cut, folded, stored, and packaged into a vesicle in the endoplasmic reticulum and golgi apparatus.

349 RNA splicing prepares pre-messenger RNA by removing introns and joining exons, which is necessary for the correct transcription of eukaryotic messenger RNA.

350 Cell structure and function are controlled through processes such as phosphylation, which can activate and deactivate enzyme functions.

EXISTENCE:

Cells and Self-Replication

351 Cell formation and self-replication like the Universe moved from smaller, simpler to larger and more complex and from the inorganic to the organic.

352 Pre-biotic/biotic transit is not lineal. Cell-like membranes and self-replication exist in pre-biotic entities, and there is life without them, i.e., viruses.

353 Some biotic entities produce or secrete mineral substances and pre-biotic material that can facilitate life, its reproduction, defense. Your fingernails are an example.

354 Once called "nanobacteria," Calcifying Nano Particles (CNP) (-200nm) are biomorphic, cell bound, self-replicating, mineral-fetuin colonies.

355 The fetuin carries material for kidney stones and plaque in the blood. The mineral forms metilites, which are aragonite calcium carbonates.

356 Aragonites form stalactites in caves and sea bottoms. Biologically controlled they form coral endoskeletons, mollusk shells, pearls, and egg shells.

357 Nanobes are ultrasmall (-20nm) underground filaments, from as deep as 5 kilometers. A similar fossil was found on a Martian origin asteroid.

358 Nanobes are cell bound and self-replicating but lack phosphate and thus RNA/DNA. They are either crystals or an entirely new life form, protobiont.

359 Viruses are life without cell membranes that use a host to reproduce. In synctia and coenocytes, multiple nuclei are in one cytoplasm, as in filamentous fungi.

360 Life at the borderline is more than cell morphology and self-replication. It is additionally the combination of structure, genetic information, and energy.

EXISTENCE:

ATP, Energy Transportation, and Enzymatic Functions

361 Adenosine Triphosphate (ATP) is a nucleoside triphosphate molecule that acts as a coenzyme for the transport of chemical energy within and between cells.

362 ATP is a coenzyme used by structural proteins and other coenzymes for biosynthesis, including DNA, motility, and cell division.

363 ATP is produced by phosphorylation (substrate-level, oxidative, and photo), cellular respiration, photosynthesis, and fermentation in cellular metabolism.

364 The many enzymes for ATP biosynthesis include ATP synthase, adenosine diphosphate (ADP), and adenosine monophosphate (AMP)and several phosphate groups.

365 ATP has a purine base (adenine) attached to ribose, the same pentose sugar as RNA, with three phosphate groups attached to the '5 carbon of the ribose.

366 When ATP is used in DNA synthesis, the ribose sugar is first converted to deoxyribose sugar by the ribonucleotide reductase. This becomes an endless cycle upon use.

367 Upon use ATP phosphate groups are freed, and ATP reverts to ADP and AMP, to be again biosynthesized from them and the phosphates in a rapid, endless cycle.

368 *Homo sapiens sapiens* organisms contain 250 grams (8.8 ounces) of ATP, but it daily biosynthesizes its body weight in ATP, indicating the speed of the endless ATP cycle.

369 The relation of ATP to AMP is the information that signals cells their energy level and controls the metabolic processes that produce and consume ATP.

370 ATP is combined with nucleic acids by polymerases in the process of transcription of the RNA and DNA genomes. It is thus a critical enzyme in the EQUILIBRA of life.

EXISTENCE:

RNA and DNA Replication and Enzymatic Functions

371 RNA is more unstable than DNA in that it contains a hydroxyl group attached to the 2' pentose ring with a risk of hydrolysis that DNA does not have.

372 RNA replication through RNA polymerase translates an RNA genome to messenger RNA (mRNA) that replicates them directly into duplicate RNA.

373 RNA replication takes place in RNA-dependent viruses by RNA-dependent polymerase that transcribes the RNA chain into a new one via mRNA.

374 RNA replication through transfer (tRNA) copies DNA code into DNA for transmission by mRNA and replication by a ribosome, which in turn is RNA.

375 RNA replication has more mutations than DNA in that the DNA double helix is a double check proofreader. RNA is usually a single, shorter strand of nucleotides.

376 In addition to coding, decoding, regulation, and expression of genes, RNA senses cellular signals and communicates responses.

377 When an RNA strand doubles up on itself, it can form by complementary base pairing groups of shorter double helices capable of chemical catalysis of enzymes.

378 In protein synthesis (translation) mRNA carries DNA information, while tRNA transfers amino acids for ribosome processing.

379 Earliest life used RNA for reproduction, protein production, and enzymes. RNA-DNA-protein life evolved because of the stability of DNA and the variety of protein enzymes.

380 Structure, RNA/DNA information/protein production, and RNA/protein self-folding to produce enzymes and ATP for energy transfer combined to produce live communities.

EXISTENCE:

Last Common Ancestor Community (LCAC)

381 The sun and solar system were formed 4.6 billion years ago (bya). The moon was formed by the Theia impact 4.5 bya with heavy bombardment until 3.9 bya.

382 From 3.8 to 3.5 bya the Last Common Ancestor Community (LCAC) of species/individuals combined elements/macromolecules/structure/energy/information.

383 The LCAC with ester lipid membranes were extremophile, hipertermophile, and independent from light but not from CO2 and other inorganic nutrients.

384 The LCAC habitat had water/high salinity/high temperature/high pressure where hydrogen sulfate/carbon dioxide/nitrates are catalyzed by ferrous ions.

385 LCAC prospered in seabed volcanic fumaloras, creating volatility/mobility that with lateral gene transfers led to development/differentiation.

386 The chemical mix of the LCAC environment facilitated the production of amino acids, polypeptides, and enzymes though RNA first and then DNA and proteins.

387 The LCAC was composed of primitive cells without a nucleus or organelles and few genes, lipid membranes with frequent lateral gene transfer that led to diversity.

388 The LCAC had characteristics similar to archaea, which differentiated first and is the most direct and oldest descendent of LCAC, followed by bacteria.

389 The virus community differentiated with parasitical activity on archaea and bacteria. They may have had cells they abandoned with the parasitical adaptation.

390 Aerobic eukaryotes differentiated over a billion years later after the Great Oxygenation created in large part by polykarotes, especially cyanobacteria.

EXISTENCE:

Earliest Single-Cell Life

391 LCAC and archaea lack identifiable fossils, but specific lipids in rock with isotopes of 3.8 bya are found in Greenland, as is biogenic graphite from 3.7 bya.

392 The oldest fossil evidence is from microbial mat fossils from 3.5 bya in Australia. These expanded with cyanobacteria, blue-green algae, peaking at 1.4 bya.

393 Earliest life consisted of one-cell microorganisms with no nuclei but rather nucleic acid in cytoplasm and reproduction by cell division (meiosis).

394 Single-cell polykarotes, archeae, and bacteria, as well as viruses, were the only life for 2 billion years until the appearance of single-cell eukaryotes 1.8 bya.

395 Eukaryotes contain a nucleus with DNA and membrane-bound organelles, including mitochondria, with its own genome, chloroplasts, and endoplasmic reticulum (ER).

396 Mitochondria derive from anaerobic cells ingesting an indigestible aerobic and establishing a symbiotic relationship based on nutrients for aerobic ATP.

397 A cell captured a photosynthesizing cyanobacteria that evolved into chloroplast for plant photosynthetic energy stored in ATP, fatty/amino acid synthesis.

398 Mitochondria produce ATP and regulate the cellular cycle: growth/differentiation/death. With ER it forms the Mitochondria-Associated ER Membrane (MEM).

399 MEM contributes to cellular physiology and homeostasis by enzymes for lipid biosynthesis and transfer and calcium signaling and transfer.

400 Mitochondria and MEM increased adaptive capacity, as did the addition of sexual reproduction (mitosis) in addition to cell division (meiosis) from 1.2 bya.

EXISTENCE:

Multicellular Life

401 All eukaryotes were single celled from 1.8 bya to 1.2 bya when simple multicellular organisms, such as red algae/coral evolved, with sexual reproduction.

402 Between 1.8 bya to 560 mya, multicellular plankton, yellow-green algae, giardia lambia, amoeba, kelp, and fungi (yeasts, molds, mushrooms) evolved.

403 The first complex multicellular organisms appeared after the Cryogenian period glaciation in the form of the Edicaran biota of disc/tubular/mollusk-like beings.

404 The Edicaran biota was short-lived (635-542 mya) and completely replaced by the Cambrian Explosion (580-500 mya), in which most contemporary phyla appeared.

405 The Cambrian biota generated jellyfish, sponges, stony corals reproducing asexually by budding; predatory, poisonous sessil sea anemones.

406 There followed the appearance of major phyla: chordata, anthropods, echinoderms, mollusks, brachiopods, of which all were sea species.

407 After the Cambrian the phyla evolved rapidly with the first sub-phylum vertebra in agnatha (485 mya), jawless fish; hagfish and lampreys survive to date.

408 Between 485-434 mya water-born species evolved. At 434 mya green algae and fungi living in lakes moved on shore, then insects and seeds appeared (400 mya).

409 The first recorded tetrapod tracks on land date from 396 mya and terrestrially resident tetrapods from 363 mya with amphibian diversification from 340 mya.

410 Reptiles appeared at 300 mya, dinosaurs (225), mammals (200), birds (150), flowers (130), bees (100), snakes, tyrannosaurus (68), primates (55).

EXISTENCE:

Primates

411 The primate order has the Strepsirrhine suborder of lemurs, loris, galagos, and the Haplorinehaapori suborder of tarsier and simians. We of course are in the suborder that includes the simians.

412 The simians include catarrhines, Old World monkeys, including apes, humans and platarrhines, New World monkeys: marmosets, squirrels, spiders, capuchins, howlers.

413 All primates have pentdactilia (five fingers and toes) on the hands and feet, as well as opposed thumbs, but some have lost this on the feet, including humans.

414 The fingers of most primates have flat nails instead of claws, are flexible with divergence and convergence, making grasping/holding/throwing possible.

415 Some primates have plantigrade rather than digitigrade feet, which allows them to stand on their rear extremities, like chimpanzees, gorillas, humans.

416 Plantigrade feet and the development of verticality led to bipedalism in humans, which, when combined with manual dexterity, allows for carrying even when running.

417 Smell is less important than touch through pads on finger tips and vision, which is in color and binocular, allowing for greater discriminatory and depth perception.

418 Well-developed cerebral hemispheres and progressive cerebral development, especially of the cerebral cortex, move from lesser apes to gorillas and humans.

419 Gestational development of fetal nutrition and prolonged postnatal child care with parental mentoring of offspring and family, community teaching/learning.

420 The evolution of the primates has progressively equipped them for manual dexterity, visual observation, cognition, and learning at individual, family, and community levels.

EXISTENCE:

Genus *Homo*

421 Predecessor of genus *homo* was the African *autralopithecus* (4 to 2 mya), which was small (1.20 to 1.40 meters), as was its brain at 500 cc, 35% of humans, but bipedal.

422 *Australopithecus* existed from 4 to 2 mya. The first homo species *H. Habilis* dates from 2.4 mya and had a greater cranial size, 650 cc, and made and conserved tools.

423 *H. Erectus* began 1.8 mya, rose 1.80 meters, had 1,100 cc of brain, and used the Achelense toolkit: axes, knives, scrappers, perforators like *H. Ergaster* (1.75 mya).

424 *H. Antecessor* (900 tya) led to *H. Heidelbergensis* (600 tya), 1.80 meters with a 1,350 cc brain. *Homo Naledi* (335-236 kya) had a small brain with hemispheres & frontal lobe.

425 A *H. Heidelbergensis* group migrated to Europe (400 to 300 tya) and evolved to *H. Neanderthalsis* (230 tya), while the group in Africa evolved into *H. Sapiens* (300-200tya).

426 *H. Neanderthalsis* stood 1.70 meters with a 1,200 to 1,900 cc cranial capacity, can be attributed early, sophisticated cave art; became extinct 40 tya because of human food competition, assimilation, violence, maybe some cannibalism.

427 *Homo Sapiens* include Indaltu (160 tya to 15 tya), Denisovians (200 to 12 tya), hobbit-like *florensiensis* (100 to 12tya), and *Sapiens* (300-200 tya-present), sole survivor.

428 Some 70 tya small groups of *homo sapiens sapiens* left Africa and the Levante to populate Eurasia, where they interbred with Devonians and Neanderthals.

429 *Homo sapiens sapiens* mixed with Denisovians migrated to Melanesia and Australia. From 3-5% of the DNA of Melanesians/Australian Aboriginals is Denisovan.

430 In some areas 17% of Neanderthal DNA was Denisovan. Apart from assimilation, these groups succumbed because of food competition & violence by us, especially with climate change.

EXISTENCE:

Homo Sapiens Sapiens

431 *Homo Sapiens Sapiens* stands 150-190 meters on average with 1,800 cc of brain. We are the sole survivor of both genus homo and of species *homo sapiens.*

432 *Homo Sapiens Sapiens* also interbred with Neanderthals. Non-Africans have on average 2% Neanderthal DNA. Cro-Magnon were early Europeans and not another species.

433 The scientific answer as to who we are is the sub-species Homo Sapiens Sapiens, species Homo Sapiens, genus Homo, family Hominidae.

434 We are of the suborder Haplorhini, order Primates, class Mammalia, phylum Chordata, Kingdom Animalia, an integral part of life on the planet Earth.

435 Where we are located in the Supercluster Virgo, subgroup Local of over 30 galaxies, the largest being Triangulum, Andromeda and our Milky Way (one off the oldest at 3.6 bys).

436 In the spiral Milky Way our Sun is located in the Orion Arm, some 25,000 light years from the center of the galactic core & a supermassive black hole located at Sagittarius A.

437 In the solar system, we are the third rocky planet from the Sun at 147.18 million kms., Mercury & Venus being closer & Mars farther. There follow Jupiter, Saturn, Uranus, Neptune, Pluto.

438 We came from, 4.8 bys of Universe, 4.6 of Earth, 3.8 of life, 2 bys to reach a second cell, and 1.4 bys to leave the sea for land. Mammals appeared 200 mys ago, primates 55 mys ago.

439 The Genus Homo first appeared 2.6 mya (million years ago), & Homo Sapiens evolved from 300,000 to 200,000 years ago in Africa, from where we spread everywhere except Antarctica.

440 **We are species and unique individual offspring of Grandfather Singularity, Grandmother Stars, Father Son, Mother Earth, stardust to be sure, an integral part of the Universe.**

EXTINCTION:

Historical Inevitability

441 The risk humans will become extinct is 100%. The risk that you and I will die is 100%. Existence and extinction, life and death are our ultimate reality.

442 This ontology is beyond our past, present, and future control. It is objective. Subjective solutions are denial. Time is everything.

443 What we can do is seek to increase the time of species survival and individual existence, as well as construct and enhance the meaning and quality of life.

444 What we have is a window of opportunity to exist and live to the fullest the time we can survive as a species and live as individuals constructing purpose beyond survival.

445 The time we can survive allows us to realize and expand human potential for safety, wellness, well-being, dignity, freedom, and purposive goal achievement.

446 Extinction is the end of existence of a species or taxonomic family when the last individuals disappears. It is the ultimate threat. It is also inevitable.

447 An estimated 99.9% of all species that have ever existed are extinct. The typical species is in existence some 10 million years before extinction.

448 Some species last hundreds of millions of years virtually unchanged. Modern crocodiles and sharks have existed for 145 million years.

449 Pseudo extinction is when a parent species disappears but has evolved into a daughter species that carries on most of the parent's genetic information.

450 Extinctions are local in an area, in the wild when captives remain, total when worldwide, co- when the extinction of one species leads to another extinction.

EXTINCTION:

Five Historical Extinction Events

451 Since life began on planet Earth, there has existed a background extinction rate, as some species appear through speciation and others become extinct.

452 Marine fossils, the most abundant, indicate a background extinction rate of 2 to 5 taxonomic families of sea invertebrates and vertebrates every million years.

453 An extinction event (also extinction-level event, ELE) occurs when a large number of species cease to exist in a relatively short period of time.

454 The five historical extinction events occurred in the last 450 my occurred 450 mya, 375 mya, 251 mya, 200 mya, and 66 mya. The sixth is largely human caused and ongoing.

455 The two Ordovician–Silurian (0-S) extinction events (450–440 mya) killed off 27% of families, 57% of genera, and 60-70% of sea species, the 2nd-largest event.

456 Late Devonian extinction (375 mya) was a 20 my series of pulses that eliminated 19% of families, 50% of genera, and 70% of sea species.

457 The largest Permian–Triassic (P-T) (252 mya) killed 57% of families, 83% of genera, and 90-96% of species (96% of marine and 70% of terrestrials, including insects).

458 Triassic–Jurassic extinction (T-J) event (200 Mya) killed 23% of terrestrial families, 48% of genera, and 70-75% of species. Most amphibians died.

459 Cretaceous–Paleogene event K-Pg (66mya) killed 17% of families, 50% of genera, 75% of all species and all non-avian dinosaurs, leaving mammals and birds dominant.

460 The two greatest events, P-T (250 mya) and T-J (200mya), involved global warming for different reasons, methane hydrate release and volcanism, respectively: early warning.

EXTINCTION:

Causes of Historical Extinction Events

461 The O-S event (450 mya) was caused by global cooling, glaciation, and lowering sea levels when the supercontinent Godwana drifted over the South Pole.

462 The late Devonian (375 my) killed brachiopods, trilobites, and coral reef by anoxic (oxygen depletion) and Euxenic (hydrogen sulfide) caused by volcanic CO2.

463 The PT (250 mya), Great Dying, saw CO2 levels rise 2000 ppm & temperature 6C because of greenhouse methane hydrate release from seabed and permafrost that accelerated global warming.

464 In T-J (200 mya) massive volcanism released CO2/hydrogen sulfide and carbon isotope, exacerbated by dissociation of gas hydrates to create global warming.

465 The K-Pg (66 mya) was due to global cooling by an 18 month impact winter that impeded photosynthesis because of the 180 x 20 km-wide asteroid that hit Yucatan, leaving the Chicxulub crater.

466 There have been 19 lesser extinction events from as old as 2400 mya to as recently as 13,000 years ago whose causes complete the list of historical threats.

467 The Great Oxygenation 2400 mya oxidized strong methane to weaker CO2 and water creating a glaciation, "Snow Ball Earth" for 400 years and biota loss.

468 In the Pliocene-Pleistocene extinction (2 mya) a stellar explosion from Scorpius-Centaurus (380-450 light years) put radioactive Iron (Fe) in deep ocean crust.

469 In the Pleistocene (640, 74, and 13 tya), the advance and retreat of ice combined with human overhunting led to the extinction of many large mammals.

470 Stellar explosions/meteors, climate change by CO2, methane, sea temperature/level changes, acidification, anoxic/euxenic events cause historic extinctions.

EXTINCTION:

Extinctions and Historical Threats to Genus *Homo*

471 In genus *Homo* many extinctions are pseudo given daughter species continuing genetic information and increasing adaptive capacity, i.e., cranial capacity.

472 This does not exclude food competition and/or predatory practices (genocides and/or cannibalism) or variable resistance to pathogens as mechanisms.

473 Terminal extinctions in genus *Homo* include *Homo Neanderthalensis* and the *Homo Floresiensis*, who did not leave successor species that continue their lineage.

474 There have also been extinction close calls. The explosion of the Toba volcano in Indonesia 77 tya was 100 times greater than any modern eruption.

475 The Toba explosion created a caldera complex 100 x 30 km. The dense rock equivalent (DRE) of the material exploded is between 2,000-3,000 cubic kilometers.

476 Toba ash fall consisted of 800 km3 with a layer 15 cm thick over South Asia and deposits in the Indian Ocean, Arabian and South China seas, and Greenland.

477 A 10-year volcanic winter and cooling, perhaps as much as —15C, ensued with attendant extinctions. Humans were reduced to 2,000 to 20,000 individuals.

478 Genetic evidence points to a bottleneck 70 tya with humans descended from as few as 1,000-10,000 breeding pairs. Other mammals have bottlenecks in this period.

479 The small pockets of surviving population accelerated the formation of races, a process that would have otherwise taken millions of years.

480 Humans survived because of their intelligence, tools, social organisation, and culture. These are the factors that have allowed humans to adapt in their evolution.

EXTINCTION:

The Sixth Holocene or Anthropocene Extinction Event

481 The Sixth Holocene event is ongoing, anthropogenic, and directly linked to human impact on other species, the environment, habitat degradation, and climate changes.

482 Human adaptations for survival have consisted of increased transformative capacity, especially in the last 250 years, with cumulative negative consequences.

483 Our impact on the environment was minimal with the Early Paleolithic toolkit of *H. Habilis* and Middle and Late Paleolithic of Erectus, Neanderthals, and Cro-Magnon.

484 The perfecting of tools included weapons that increased destructive capacity. The Neanderthals couldn't compete with human pressure.

485 Mesolithic overhunting contributed with climate change and disease to extinctions of mammoths, saber tooth tigers, and others 10 tya, our first impact.

486 Mass deforestation was our second impact by the use of fire by Paleolithic hunters and to clear the land for Neolithic and Mesolithic agriculture/grazing.

487 Deforestation increased with civilization, cities, roads, increased populations, and the use of wood for fuel, construction, and artefacts of every kind.

488 In 500 years these processes accelerated, and today FAO estimates that 80,000 square miles, 2% of total forests, are lost yearly; only 7% of Earth is still in forest.

489 Urban sprawl, productive and infrastructure projects take forests in the North, while in the South pressure is on topical hard wood and from agricultural frontiers.

490 Deforestation habitat loss has led to extinctions and threatened species, combined with air, water, and soil pollution, intentional/unintentional invasive species.

EXTINCTION:

Population, Economic, Technological Growth

491 Human impact increased geometrically since the Industrial revolution because of the combination of population, economic, and technological growth.

492 Population increased from 375 million in 1420 to 1 billion in 1802, 2B in 1927, 3B in 1961, 4B in 1974, 5B in 1987, 6B in 1999, and 7 billion in 2013.

493 Birth rates are declining, but projections are 8.2B in 2030 and 9.1B in 2050, when the rise will crest with 2B more persons than in 2013.

494 Humans are literally crowding out other species by our sheer numbers, our increasing demand for space and resources, and contamination from our activities.

495 Economic growth has been even greater. World GDP was US$1.4T in 1960 and US$75T in 2013; in constant dollars it went from US$9.2T to US$56.5T, 6 times more.

496 From the Industrial Revolution toolkit of coal, steam, and mechanics, technology has created the nuclear and aerospace toolkits, leading to nuclear-tipped missiles, laser weapons.

497 The electronic, information, communications, and robotics toolkits, including the Internet, have revolutionized all spheres of activity.

498 Bio-technology, new materials, and nano technology are the frontier that with 5G Information and Communications Techology (ICT) will further expand transformative capability in a declining environment.

499 New technologies further expand transformative capability in a declining environment, generating employment/labor/ social /political instability.

500 Population/economic/technological growth have increased demand for resources and extractive capacity beyond water, soil, and food sustainability.

EXTINCTION:

The Degradation of the Marine Biota

501 Life began in the sea, and nowhere is the speed of degradation and decline more apparent or more massively devastating to bio diversity than in the sea.

502 Since industrial fishing began in the mid-20th century, 90% of all large fish species have been depleted, including deep sea tuna, swordfish, sailfish, and marlin.

503 Coastal overfishing includes cod, halibut, skate, and flounder. Enormous stocks such as the Grand Banks were overfished beyond recovery.

504 Overfishing includes tropical yellow fin tuna, cold water blue fin tuna, and, in between, albacore. All areas of the sea are overfished.

505 Increasing demand with decreasing resources is a terminal formula. The Law of the Sea is not fully in force because of nonacceptance by some; anarchy reigns in international waters.

506 Climate change, ocean temperature rises, and acidification are depleting coral, a breeding ground/safe haven for offspring of many species.

507 Anoxic processes (oxygen depletion) have led to 150 anoxic dead areas in the seas, also floating refuse islands, deaths of marine biota by ingesting plastics or their particles.

508 Pollution from our production and consumption, including mercury, affects fish stocks even in pristine arctic waters because of currents to seas with low evaporation rates.

509 The progressive loss of coral and of the continued depletion of large species will have unknown consequences in the marine food chain.

510 Algae, plankton, and krill are at the bottom of a food chain and under stress. They are the largest food biomass on the planet. Life began in the sea. Dead seas will presage a dead planet.

EXTINCTION:

Halocene Extinction Projections

511 Additional to the demise of the Neanderthal and the Ice Age mammals, humans hunted to extinction the Moa bird in New Zealand and the two-finger sloth in the Caribbean.

512 Leaky and Lewin consider the Holocene event the largest since the Cretaceous Paleogene (K-Pg) of 66 mya caused by the Chicxulub asteroid impact.

513 E. O. Wilson considers that in the current Holocene or Anthropocene event that half of all species are at risk because of a human-caused species holocaust.

514 Currently a species becomes extinct every 20 minutes. More than 8,000 full species are on the IUCN red list of endangered species, 1,100 of them mammals.

515 Large carnivores (lions, tigers, leopards, and cheetahs), elephants, rhinos, and gorillas have decreased by up to 90 per cent in the last 20 years.

516 UNEP estimates that 1,000 species of mammals face extinction in the next 30 years, one quarter of the total number, because of destruction of habitat.

517 Large mammal habitats are lost by human settlements in wilderness regions and rain forests, wetlands destruction, climate change, and industrial pollution.

518 Once extinctions occur, there is a loss of irreplaceable genetic material and associated bio-chemicals that form part of life support systems.

519 Biodiversity is part of the survival strategy of life on earth, and the interrelations of life support systems can have both positive and negative synergies.

520 Does anyone of us really doubt that the queue of species marching toward extinction because of our impact on the environment does not include our own?

EXTINCTION THREATS:

Cosmological Bolides

521 Extinction threats to life on Earth are anthropogenic, partly examined above, cosmological, geological/epidemiological/climatological/nutritional.

522 Cosmological threats consist of bolides, supernova explosions that can fatally affect the Earth. They are all inevitable but with thousands of years between major occurences. Solar radiation is also a threat.

523 Bolides are comets, asteroids, meteors/meteorites. Comets are low-density, irregular-shaped entities consisting of dust, ice, CO2, ammonia, methane.

524 Comets are found in the Oort Cloud beyond Pluto, and some have short/long-term orbits to the inner solar system, where they voyage leaving luminous aura and tails.

525 Comets range in size from nucleus diameters of 30 km to 100 m. Their tails contain particle debris that creates meteor showers in Earth passage.

526 Asteroids are rocky, either carbon or metallic, smaller than a planet, and larger than a meteorite. They are found between the Sun and Neptune.

527 There are dangerous NEA (Near Earth Asteroids), an Asteroid Belt between Mars and Jupiter, Troyanos in Jupiter orbit, Centauros in the outer solar system.

528 A meteorite is a 10 m or less rocky/metallic/mixed debris from solar formation/a comet/an asteroid that leaves a light trail, meteor, upon entry.

529 There is an unverified 480 km crater under Antarctic Ice caused by a bolus of at least 55 km that can be associated with the Permian Triassic extinction 252 mya.

530 The Chicxulub Crater of Yucatan (180 km from 10 km asteroid) 65 mya caused the K-T event by a multi-year atmospheric dust winter disrupting photosynthesis/world food chain.

EXTINCTION THREATS:

Cosmological, Supernova Explosions/Solar Radiation

531 A supernova is the explosion of a star, the largest such event in the Universe, more luminous than a galaxy and with more energy than an average star's lifetime.

532 Supernovae type Ia are formed by the gravitational accretion of white dwarfs in binary systems whose density detonates thermonuclear fusion.

533 Supernovae type II are 8/50 times larger than our Sun and fuse heavy elements in layers except an iron (Fe)-nickel (NI60) core that implodes/explodes.

534 IK Pegasi in Pegasus is the closest visible binary star system to Earth at 150 lyrs from our solar system but is slower, and the distance is increasing, reducing risk.

535 There appears to be no type II Supernova candidate at a dangerous distance, and the dimmer, more numerous white dwarf threats are greater but further in the future.

536 An extinction event 2.2 mya reduced sea life by 50% because of a Centaurus Supernova explosion as shown by a 60 cm layer of Fe and radioactive NI60 on the Pacific Ocean floor.

537 Supernovas leave spinning neutron stars and pulsar remnants or nebulae like the crab. With heavier density they collapse internally, forming a black-hole sink.

538 Supernova explosions threaten ozone layer degradation on Earth as gamma rays convert N to NO2; an explosion at 26l yrs from Earth would destroy half our ozone layer protection.

539 Ozone layer depletion allows more solar radiation, UCE, and UVA to penetrate to Earth, causing damage and potentially threatening life itself.

540 Apart from Supernovas, there is anthropogenic ozone depletion by Chlorofluorocarbons (CFCs) and Hydrofluorocarbons (HFCs). Montreal Protocol (1992) controls staged elimination.

EXTINCTION THREATS:

Geological

541 After intense volcanic activity in the Early/Late Bombardment periods, volcanic activity has continued through faults, particularly along tectonic plates.

542 The greatest geological threats to biodiversity are from super volcano eruptions and gases emitted by lava from large igneous provinces.

543 Supervolcano eruptions can create dust winters that can produce climate change and global cooling, disrupt photosynthesis and world food chains.

544 A Volcanic Eruption Index (VEI) has 8 categories based on the amount of material erupted. VEI 8 consists of eruptions expelling more than 1,000km3 of material.

545 The latest supervolcano (VEI8) was Lake Taupo in New Zealand 25.5 tya that expelled 1,050 km3. The second latest in Lake Toba, in Indonesia 74tya, expelled 2,500km3.

546 Lake Taupo and Lake Tabo are the only VE-8 during the period of *H. Sapiens*, and Tabo created a near human extinction and genetic bottleneck.

547 The largest explosion was the Garita Caldera in Colorado USA 27.8 mya expelling 5,000km3. Unconfirmed is Paraná and Etendeka in the Phanerozoic with 15,000km3.

548 By comparison, Mount Pinatubo 1991 had 10km3 of ejecta, Mount St. Helens 1980 (1.2 km3), and Mount Tambora 1815 (160km3), making 1816 "the year without a summer."

549 The Siberian Traps flood basalt 250 mya (1-4km3 of lava covering 7Mkm2, the world's largest) coincided with and contributed to the Permian-Triassic (P-T) event.

550 Volcanic event threats to species are usually localized but super volcanoes and flood basalt gases do have the potential to pose universal extinction risks.

EXTINCTION THREATS:

Epidemiological

551 Pandemics are worldwide, epidemics one area, infectious diseases are battles between pathogens and immune systems, conditioned by nutrition and health levels with vaccines as a defense.

552 Historically, populations were decimated by typhus, measles, smallpox, leprosy, bubonic plague, cholera, tuberculosis, and influenza from ancient times.

553 Population losses of 50%+: Athens (430BC, typhoid), "Justinian" Europe (550-700, bubonic), England (1361-1480, bubonic in waves).

554 New European diseases in the Americas killed 95% of the population and in the Pacific, Easter Island 90%, Australia 50%, Fijians 33%, Hawaii 25%, while syphilis went to Europe.

555 Typhus "camp fever" caused higher mortalities than combat, 19th-century cholera killed tens of millions, and the Spanish flu 50 million in 1918-1919.

556 Currently the highest annual world infectious mortality is from influenza and pneumonia 3,5 M, HIV/AIDS 1.8 M, Malaria 825 K, Meningitis 340 K, Hepatitis B 128 K, and COVID-19 with a low death rate and high fatality numbers.

557 Current challenges and epidemic risks include malaria, dengue, chicunkuya, zika, HIV/AIDS, tuberculosis, meningitis, NIPA, Ebola & Coronaviruses: MERS, SARS, and Novel Coronavirus (SARS-CoV-2) / COVID-19.

558 The greatest current threats are rapidly mutating viruses, including animal-to-person contact, person-to-person, contact-to-airborne, and increasing resistance to medicines.

559 With regard to the humans vs. viruses, vaccinations/ antibiotics/other medicines, treatment protocols, the balance improved for us, but new viruses, mutations, and antibiotic resistance represent a viral counteroffensive.

560 The Novel Coronavirus (SARS-CoV-2) and COVID-19 demonstrate the damage a new virus can do to billions, especially the poor; public health, health workers, health care systems; businesses, jobs, markets, economies.

EXTINCTION THREATS:

Anthropogenic, Nuclear Fission

561 The atom was split in the laboratory of the German physicists Otto Hahn and Fritz Strassmann in 1938 and was interpreted in 1939 by Lise Meitner and Otto Frisch in 1939.

562 A neutron is absorbed by Uranium 235 transformed into Uranium 236, splits into Barium 141 and Krypton 91, freeing neutrons/gamma rays and electromagnetic/radiation/kinetic energy.

563 Frisch denominated this process of nuclear transmutation as "nuclear fission" by analogy to the biological separation of living cells.

564 The German Wehrmacht began a nuclear research program in 1939, "Uranverein," weakened by a migration of physicists for political reasons and the drafting of scientists and engineers.

565 Political interference came from the SS and Deutsch Physik movement, which attacked the teaching of "Jewish physics," as well as bureaucratic instability in the army and universities.

566 The German nuclear program had three pillars: uranium and heavy water production (the latter sabotaged and bombed in Norway), nuclear reactors, and uranium isotopes.

567 By 1942 the Wehrmacht realized that Germany did not have sufficient industrial capacity to produce a bomb and dedicated its remaining 1,200 tons of uranium to solid-core ammunition.

568 Hungarian physicists, refugees from the Nazis, Leo Szilard and Eugene Wigner wrote in a 1939 letter to President Roosevelt, signed by Albert Einstein, expressing they feared a Nazi bomb.

569 Roosevelt requested a report on the Szilard-Einstein letter that in December, 1939, found uranium would provide a possible source of bombs with vastly superior destructiveness.

570 In Britain, the 1939 Frisch-Peieris memorandum established the critical mass of Uranium 235 for a bomb at a manageable 10 kilos, and the Maud Committee began a nuclear weapons program.

EXTINCTION THREATS:

Anthropogenic, Manhattan Project

571 The precursors of the Manhattan Project began modestly and expanded enormously after Pearl Harbor and the entrance of the U.S. in the war against the axis.

572 The project reached 24,000 skilled employees in 30 sites and spent US$2 billion (equivalent to US$23.5 billion in 2016). Robert Oppenheimer headed the project with Gen. Leslie Grove.

573 The project absorbed the British program, Tube Alloys, and the Canadian program. British scientists, including Karl Fuchs, joined; there were Canadian research sites.

574 Three approaches were pursued for isotope separation: electromagnetic (U.C. Berkeley), thermal (Carnegie), and gaseous (Columbia U.), which came to predominate.

575 In February, 1941, Glenn Seaborg at U.C. Berkeley bombarded uranium with deuterons, creating neptunium (element 93) that beta decayed into a new fissile element plutonium (94).

576 In December, 1942, Enrico Fermi of the University of Chicago produced the first engineered, self-sustaining nuclear chain reaction when the nuclear reactor Chicago Pile-1 went critical.

577 In a typical nuclear chain reaction, a neutron bombards Uranium 235, releasing energy and neutrons, one absorbed by stable uranium 238/one not absorbed/one again by a U235 atom.

578 Criticality is reached when reproduction of neutrons is stable across generations, creating a chain reaction, while there is fuel in a controlled reactor or an explosive bomb.

579 The Trinity Test of the "Gadget," an implosion-design plutonium "Fat Man" bomb produced at Los Alamos on 16 July, 1945, yielding 22 ktons of TNT, began the nuclear age.

580 One of the two greatest scientific achievements in history, the harnessing of the atom (the other being the decoding of DNA), brought us a man-made threat to existence.

EXTINCTION THREATS:

Anthropogenic: Nuclear Warfare

581 The war in Europe ended on May 8, 1945 & the new weapon threatened Japan. A group of the scientists involved in making the bomb had doubts & moral objections about mass destruction.

582 On June 1, 1945, the Interim Committee on Nuclear Policy advised against a demonstration to the Japanese of the power of the bomb but instead argued for a surprise combat attack.

583 Arthur Compton offered scientists with concerns the opportunity to present them to the Interim Committee. The result was a report by Franck, Rabinowitz, and Szilard.

584 The Franck Report urged a desert island demonstration and international control of nuclear weapons to avoid an inevitable arms race and the threat of mutual destruction.

585 In July, 1945, a group of 70 Manhattan Project scientists signed the Szilárd Petition requesting a demonstration of atomic power before use on people and international control.

586 Szilard held nuclear weapons were designed to destroy cities, military & civilian targets, with moral/legal implications, but the U.S. was already involved in same.

587 In January, 1944, General Curtis LeMay abandoned imprecise high-level bombing and precise low level bombing with anti-aircraft losses in Japan for incendiary carpet bombing of 67 "paper cities."

588 The incendiary carpet bombing of Tokyo,8-9 March, by 337 B-29's (27 lost) killed 97,000, wounded 140,000, destroyed 41 square km and 286,000 buildings and left one million homeless.

589 Three weeks after Trinity, 6 August 1945, Hiroshima was bombed with a Little Boy uranium 235 gun type bomb (15 kt), killing 146,000 people from the blast, radiation, disease.

590 Three days after Hiroshima, 9 August 1945, Nagasaki was hit with the Fat Man plutonium implosion bomb (21 kt), killing a total of 80,000 people, 226,000 between the two cities.

EXTINCTION THREATS:

Post-War, Nuclear Espionage, and Capture

591 After U.S. fire/nuclear bombings and the Soviet entry into Manchuria, Emperor Hirohito read a radio address (15 August 1945) announcing acceptance of unconditional surrender.

592 The surrender instrument was signed 2 September 1945 aboard the *USS Missouri* in Tokyo Harbor in accordance with the Potsdam Declaration calling for unconditional surrender.

593 Many Manhattan scientists, including Oppenheimer, were outraged at the second attack on Nagasaki, considering it unnecessary as Hiroshima was sufficient demonstration.

594 Einstein and Szilárd organized the Emergency Committee of Atomic Scientists in 1946 to promote world peace, no use of nuclear weapons, opposition to H bomb, promotion of N energy.

595 The wartime U.S. Alsos Mission spied on German science programs, sought to capture resources & scientists & avoid Soviet captures, both behind lines and in occupied areas.

596 Alsos, Epsilon, and the postwar Overcast and Paperclip relocated 1600-plus German Scientists in the U.S., especially rocket specialists (Werner von Braun headed U.S. program).

597 The Soviet Union penetrated Manhattan with multiple spies, the most important being Karl Fuchs, a lead physicist with both access and understanding of the information.

598 Karl Fuchs informed on Tellers super bomb concept, Theodore Hall on Fat Man design, others on bomb triggers, gaseous separation of isotopes, plutonium fuel, lithium.

599 Karl Fuchs was imprisoned & relocated in Dresden, East Germany; Morris Cohen and his wife were freed in USSR prisoner exchange; George Kovall was named Hero of Russia (2007)

600 David Greenglass implicated his sister & brother-in-law, Julius & Ethel Rosenberg, both executed, with testimony partly designed to protect his wife, and served 10 years of 15 years.

EXTINCTION THREATS:

Anthropogenic: U.S. Thermonuclear Program

601 The West expected a Soviet bomb 1953/1954 and was surprised by RDS-1(First Lightning), an implosion plutonium "fat man" bomb(22ktn), in Semipalatinsk, Kazakh SSR, on 29 August 1949.

602 After the Soviet atomic test, there began a second race to develop the thermonuclear bomb that Edward Teller had advocated in 1941. Teller and Ulam completed a design in 1952.

603 The Teller-Ulam design: a fission bomb compresses a fusion fuel secondary stage causing a fusion reaction within a uranium-238 tamper and around a plutonium sparkplug.

604 A radiation reflecting case contains the X-rays from fission, augmenting the explosive power of the fusion of the hydrogen isotope fuels (tritium/deuterium/lithium deuteride).

605 The basic concept of fusion was tested in the George Burn at Enewetak Atoll, Marshall Islands, on 8 May 1951, when a fission bomb fused deuterium and tritium (225 kilotons).

606 In the George Test fast neutrons from the fusion caused fission in Uranium-238, defining the tamper, and the radiation implosion trigger was successful, permitting final design.

607 The first thermonuclear explosion,"Ivy Mike," took place at Bikini Atoll on 1 November 1952 with a yield of 10.4 megatons (450 x Nagasaki) using impractical liquid deuteride.

608 A solid lithium deuteride fusion fuel test “Castle Bravo” on 1 March 1954 at Enewetak yielded 15 megatons, 2.5 times the programmed, widely spreading unexpected fallout.

609 Development in the 1950-1980’s consisted of smaller weapons to fit missiles, submarine-launched missiles, and multiple independent targetable reentry vehicles (MIRVs).

610 The U.S. produced 70,000-plus nuclear warheads, more than all others (cost US$9.08 trillion), including delivery systems; currently 7,100 warheads, 1,890 deliverable (1430 megatons.)

EXTINCTION THREATS:

Anthropogenic: Soviet Nuclear Program

611 Georgy Flyorov wrote Stalin in April, 1942, that German, British, and U.S. nuclear physicists had stopped writing in physics journals, indicating involvement in secret programs.

612 The Soviet program was ordered by Stalin in 1942, slowed by German invasion, accelerated by Hiroshima, and aided by captured German scientists and successful Manhattan espionage.

613 The Soviet Operation Osoaviakhim transferred 2,200 German specialists and their families, test centers, factories, and equipment to the USSR from the Soviet-occupied zone.

614 Information from Fuchs, Morris Cohen, David Greenglass, George Koval, and Theodore Hall saved false starts, oriented research, and cut development time in the Soviet nuclear program.

615 Intelligence and German scientists complemented the work of Soviet scientists led by Flyorov, Sakharov, Kurchatov, and Davidenko, who developed the H-bomb in three phases.

616 The first phase was work on a "super bomb" with an atomic bomb triggering fusion in liquid deuterium, Turbá, until 1953; Teller had realized in 1951 it would not have sufficient heat.

617 The second phase, sloika, consisted of alternating layers of fission-fusion-fission with unenriched uranium and deuterium (RDS-6s), tested 12 August 1953 (400 kilotons).

618 "Sarjov's Third Idea" was for a two-stage fission-fusion ignited by the A bombs X-ray radiation, RDS-37, tested 22 November 1955, 1.6 megatons (100x1949), at Semipalatinsk.

619 The Tsar Bomba (RDS-220) in Novaya Zemlya, 30 October 1961, with capability of 100 megatons reduced to 50 (so as not to permanently damage the site) was the largest explosion ever.

620 The Soviet Union/Russian Federation have produced 55,000 nuclear warheads with a peak stockpile of 45,000 warheads in 1988; it currently has 7,300, 1,444 operational (801.5 MTs).

EXTINCTION THREATS:

Anthropogenic, UK, and French Nuclear Programs

621 A memo by German refugees Rudolf Peieris & Otto Frisch in March, 1940, argued that the Uranium-235 could produce a huge bomb, leading to UK MAUD Committee study and the Tube Alloy.

622 The Tube Alloy project worked with UK and Canadian researchers, as well as the Dane Nils Bohr. It was merged into the Manhattan Project by the Quebec Agreement of 1943.

623 When the McMahon Act of August, 1946, forbad U.S. sharing of nuclear information, U.K. leaders decided on their own bomb to maintain world power status & dignity in relation with U.S.

624 A Fat Man implosion plutonium bomb tested 3 October, 1952, in "Operation Hurricane" on Monte Bello islands of Australia was followed by H-bombs in Operation Grapple in 1957-1958.

625 Thermonuclear U.K., in Sputnik context, was readmitted to the "special relationship" by the 1958 US-UK Mutual Defense Agreement and receipt of a MK28 warhead for manufacture.

626 The U.K. produced 835 nuclear warheads, peak stockpile of 520 (1970's) for V bombers. Currently 215 warheads, 120 usable on Trident II nuclear submarine missiles (12 megatons).

627 French scientists were pioneers in the study of radiation, Marie Curie and her assistant Bertrand Goldschmidt, who invented the plutonium extraction method while working on Manhattan.

628 The Alsos Mission undertook Operations Big and Harborage to capture German scientists and assets in SW Germany before the French Army took control of their occupation zone.

629 De Gaulle approved Force de Frappe free of NATO, achieved with Gerboise Bleue fission (Algeria, 1960, 70tn); Canopus fusion (Polynesia, 1968, 2.6Mtns); Israel collaborated.

630 France has made 1,100 warheads, peak stockpile of 540 (1992); currently has 300 warheads, 290 usable on bombers and submarine-launched ballistic missiles (SLBMs), 51.6 megatons.

EXTINCTION THREATS:

Anthropogenic: Chinese and Israeli Nuclear Programs

631 Mao Tse Tung ordered the Chinese nuclear program in 1954. Uranium enrichment and plutonium plants and a test site were created with support of the Soviet Union until the 1960 rift.

632 The first nuclear test was "496," 16 October 1964, and the first thermonuclear test was 17 June 1967, both undertaken at the Lop Nur Nuclear Weapons Test Base, Malan, Xinjiang.

633 China built 600 warheads since 1964, peak stockpile of 434. Largest test was 4Mt (1976). A.K. Khan's centrifuges assisted China, who assisted Pakistan with material and designs.

634 The last Chinese test was in 1996, and China prides itself on having the smallest arsenal of major powers for a "minimum deterrent" combined with a pledge of "no first use."

635 China has 260 warheads for 294 megatons with 478 delivery devices between 76 ICBMs, 108 MRBMs, 204 SRBMs, 54 cruise missiles, 36 SLBMs, and 300 bombers (Sukhoi Su-30/H6s, JH-7).

636 Israel doesn't confirm nuclear weapons, but the South Atlantic Flash seen 22 September 1979 by the U.S. Vela satellite is held to have been a South African-Israeli test.

637 Israel was helped by France through the sale of a reactor for the Dimona Nuclear Research Center. Whistleblower Mordechai Vanunu revealed the secret Israeli nuclear program.

638 Israel has between 75 and 400 warheads for a classified megatonnage; delivery by ICBM, mid-range, submarine-launched ballistic missile (SLBM), F-15/F-16 fighter-bombers.

639 According to Mordechai Vanunu, Israel has mass produced neutron bombs. China and France have tested but didn't deploy. The U.S. deployed them in the 1980s but later retired them.

640 Israel was against the Iran nuclear agreement, celebrated U.S. President Trump's withdrawal, and threatens war on Iran, not excluding a nuclear attack, if it acquires nuclear weapons.

EXTINCTION THREATS:

Anthropogenic: India, Pakistan, and Saudi Arabia

641 H.J. Bhabha founded the Institute for Fundamental Research in Trombay (1944), received Canadian CIRUS research reactor 1954, which produced plutonium, as did Project Phoenix 1958.

642 "Smiling Buddha" underground fission test of implosion "Fat Man" at Pokran, Rajastan (8kt.) May 4, 1974, and fusion test Operation Shakti I or Pokran-II (45kt.) 11 May 1998.

643 Three weeks after Pokran-II, 28 May 1998, Pakistan undertook the Chagai-I tests of 5 underground fission weapons (40kt.) in the Ras Koh hills of Balochistan.

644 Two days later in the Kharan Desert of Balochistan, there was the testing of an implosion-type boosted fission plutonium bomb (20kt). Neither India nor Pakistan have tested after May, 1998.

645 Munir Ahmed Khan of the IAEA later headed research reactors; AQ Khan of URENCO of Holland brought an enrichment technology to Pakistan, forming Kahutla Research Laboratories.

646 China assisted Pakistani and AQ Khan helped China in enrichment, as well as Libya/Iran/North Korea. U.S. wanted to extradite Khan but Pakistan protected national hero.

647 India has a no-first-strike and minimum deterrence policy. Pakistani policy is asymmetrical; given conventional force disadvantage, it would attack India first with nuclear weapons.

648 India has produced 120 warheads, all in stock, while it has the Agni V missile (5000kms) among 68-plus land- and submarine-based missiles, 48 aircraft (Mirage 2000s and SEPECAT Jaguars).

649 Pakistan has produced 130 warheads, all in stock, with Shaheen missiles (2750kms), Babur cruise missiles (700-1000kms), 36 aircraft (F-16s and Mirages).

650 Saudi Arabia would acquire Pakistani nuclear weapons if Iran achieves them. In 1987 the Saudis acquired 36-60 nuclear-capable Chinese DF-3a missiles (4-5000kms) for 3 50-100kts.

EXTINCTION THREATS:

Anthropogenic: North Korean and Libyan Nuclear Programs

651 In the 1980s North Korea began uranium fabrication and conversion, first nuclear test of -1kt (2006), by 2-7 kilotons (2009), 6-9 ktns (2013), 14 ktns (02-2016), 25 ktns (09-2016).

652 U.S. and six parties (N. Korea, S. Korea, China, Russia, Japan, USA), IAEA, United Nations and actions have offered possible settlements but no definitive results.

653 North Korea has 13-60 nuclear weapons and the Hwasong-15 ICBM (13,000 kilometers). The sixth test of a "hydrogen bomb" was a 250-kiloton blast producing a 6.1 earthquake.

654 The arrival of North Korea as a nuclear power has caught the attention of the United States and has gone from threats to negotiations on denuclearization of the Korean peninsula.

655 Libya sought assistance from the Swiss nuclear scientist Friedrich Timmer (1980s) on centrifuges; also bought 1,200 tons of unreported uranium from French mines in Niger.

656 The Soviet Union built a 10 MW nuclear reactor for the Tajoura Nuclear Research Facility under IAEA safeguards, but conversion research was undertaken with foreign support.

657 Libya sought expertise from Bhutto's Pakistan, China, North Korea, ex-Soviet scientists, & AQ Khan network, but key centrifuge shipment was captured at sea by U.S. in 10-2003.

658 Upon renouncing its still rudimentary nuclear program in December, 2003, Libya with U.S. assistance and IAEA supervision turned over equipment, materials, and documents.

659 Despite normalization of relations with the U.S. and Europe, a UN Security Resolution to protect Benghazi was used by France, U.K., & the U.S. to give bombing support to rebels.

660 Air support for rebels led to regime change and the murder of Ghaddafi. Contrast of nuclear-less Ghaddafi (or Hussein) and nuclear Kim in North Korea weakens non-proliferation.

EXTINCTION THREATS:

Iran and Nuclear Withdrawal Countries

661 In the 1980s Iran received assistance from the Soviet Union and China in uranium mining/milling/conversion/reactors. IAEA has indicated no evidence of a nuclear weapons program.

662 Iran researched (1990s) conversion, plutonium separation, and enrichment, receiving gas centrifuge support from the AQ Khan network; Natanz facility reached 19,000 centrifuges.

663 Natanz was secret until outed by Iranian exiles in 2002. In 2012 the 16 U.S. intelligence agencies declared that Iran continued enrichment but had no nuclear weapons program.

664 After 20 months of negotiations (2015), the Joint Comprehensive Plan of Action (JCPOA) was signed between Iran and the P-5+1 (USA, France, UK, Russia, China + Germany).

665 Under the agreement Iran is constrained for 15 years to only use first-generation centrifuges and its stockpile of low-enriched uranium reduced by 97%, eliminating medium enriched.

666 With JCPOA some sanctions against Iran were lifted. On 8 May 2018 the U.S. announced withdrawal and with Israel applied sanctions. EU resisted and sought to create new payment system.

667 U.S. maximum pressure on Iran, including oil & financial measures, as well as assassination of Soleimani, Iran renewed nuclear program while Europe demanded compliance.

668 With the ratification of the Lisbon Protocol to the Non-Proliferation Treaty (1994), Ukraine, Belarus, and Kazakhstan agreed to transfer to Russia or destroy nuclear arms (1996).

669 South Africa developed seven gun-type fission bombs in the '70s and '80s with Israeli support, tested in Vela Incident over South Atlantic 1977. Weapons dismantled 1989.

670 Belgium, Germany, Italy, Netherlands, and Turkey host NATO warheads. Algeria, Argentina, Brazil, Iraq, Italy, Sweden, Switzerland, Syria, Taiwan discontinued weapons programs.

EXTINCTION THREATS:

Nuclear Tests, Contamination, and Treaties

671 There have been 1,892-plus nuclear tests (635Mts), 86% from the USA/USSR. There were 520 atmospheric and under-water nuclear tests (545 Mts, 1945-92) and 1,352 underground (90Mts).

672 The U.S. had 1,054 tests (1149 devices, 219 atmospheric), the USSR 715(969 devices), France 210 (50 atmos.), UK (45), China 45 (23 atmos.), India 6, Pakistan 6, North Korea 7.

673 The peak was 1961-1962 with the Soviet Tsar Bomba (50Mts.) and U.S. Dominic series (31 tests, 38.2 Mts), including submarine-launched and missile-launched high-altitude nuclear tests.

674 First major contamination was the U.S. test Castle Bravo (15mt., 1954), which exceeded expectations, and spread radiation over Marshall Islanders and on a Japanese fishing boat.

675 Higher yields and wind changes also affected the Nevada test site (904 detonations) with radiation dispersion to Nevada, Utah, Idaho, Montana, the Great Plains, Missouri, others.

676 Increased fallout led to radiation sickness, thyroid cancer due to Iodine 131, and birth defects. The U.S. Army and Marines competed over nearest proximity of soldiers to blasts.

677 The National Cancer Institute estimates 11,000 excess deaths due to fallout. A Compensation Law (1990) of US$1.38 billion for Pacific and U.S. test site and wind-blown radiation.

678 The Partial Test Ban Treaty (1963) forbad atmospheric tests. France (1974) and China (1980) stopped later, but neither they nor North Korea signed.

679 The Comprehensive Test Ban Treaty (CTBT) in 1996 forbad underground tests; 183 countries have signed; eight Annex II countries have not ratified, including the United States.

680 The motivation of the partial and comprehensive test ban treaties was to eliminate the arms race; U.S. out because of possible future nuclear arms modernization tests.

EXTINCTION THREATS:

Nuclear Arms Races, ABN, SALT, and ITN

681 The Treaty on the Non-Proliferation of Nuclear Weapons (NPT, 1970, extended 2015) restricts nuclear arms to the UN Security Council 5 (US/Russian Fed./China/UK/France).

682 The NPT, in addition to legalizing nuclear arms for the UN Security Council 5, seeks to avoid proliferation, promote peaceful uses, and lead to total nuclear disarmament.

683 There are 191 signatories of NPT, the most of any treaty. The limitations of the NPT are evident in that N. Korea, Israel, India, and Pakistan and South Sudan are outside NPT.

684 Deterrence from the use of nuclear weapons during the cold war was based on the concept of Mutually Assured Destruction (MAD). Overproduction of weapons led to MAD "overkill."

685 The expense and madness of overkill led to nuclear arms treaties for specific weapons. The US-USSR Strategic Arms Limitation Talks (SALT I, 1972) limited ICBMs, MIRVs, SLBMs.

686 SALT I facilitated the Anti-Ballistic Missile Treaty (ABN, 1972), which limited ICBM sites with ABN protection to one each to avoid an ABM arms race and first-strike capability.

687 The US-USSR SALT II (1979) limited long-range and MIRVed missiles to 1,320 each and banned new missile programs. The US withdrew after the Soviet entry into Afghanistan.

688 The US-USSR Intermediate-Range Nuclear Forces Treaty (INF,1987) banned short- and medium-range land-based missiles (500-5500kms). By May, 1991, 2692 missiles were eliminated.

689 The US-USSR Strategic Arms Reduction Talks I (START I, signed 1991, effective 1994, expired 2009) limited each side to 6000 warheads on 1600 ICBMs and strategic bombers: 80% less.

690 START II (signed 1993, U.S. ratified 1996, Russia 2000) banned multiple independently targetable reentry vehicles MIRVs, fell apart when U.S. left ABM treaty (2002).

EXTINCTION THREATS:

Nuclear Arms Treaties, Star Wars

691 Start III (1997) sought further warhead and delivery vehicle reductions. Never concluded because of Russian opposition to U.S. bombings (Serbia), eastward expansion NATO, and missile defenses.

692 The US-Russian Federation Strategic Offensive Reductions Treaty (SORT, 2003-2011), Treaty of Moscow, limited both parties to 1700-2200 operational warheads.

693 The US-Russian Federation New Start Treaty (signed 2010, effective 2011) limited each side's deployed warheads and bombers to 700 and deployed warheads to 1550, -66% from START I.

694 By February 2018 the USA had 652 deployed intercontinental (ICBMs)and submarine-launched (SLBMs) ballistic missiles and heavy bombers with 1350 warheads; Russian had 527 with 1,444.

695 The US-USSR arm race peaked with "Star Wars," the Strategic Defense Initiative (SDI, 1983), which ended the ABN ban and weaponized cost as Soviets couldn't match U.S. spending.

696 The SDI consisted of scientific and technological studies in national laboratories, universities, and industry on high-energy physics, super-computation, advanced materials.

697 SDI, the Ballistic Missile Defense Organization (BMDO, 1993), and Missile Defense Agency (MDI, 2002): boost, ascent, mid-course, terminal phases: Patriot/THAAD/Aegis/Arrow/MEADS.

698 The U.S. withdrew from ABM treaty (2002), raising first-strike capabilities and fears. The Russian S-400 Triumf system (2007) has four tiers, 400, 250, 120, and 40 kms. ABMs.

699 U.S. nuclear modernization (2017-2046, cost US$1.7 trillion) and new Russian technologies (ex. Vertical missile, hydrogen bomb drone torpedo 6000 km. range) augur new arms race.

700 The purpose of nuclear modernization is smaller and hence more usable weapons, but they will still be larger than the Hiroshima weapon, making us less rather than more secure.

EXTINCTION THREATS:

Nuclear Attack[1]

701 The threat of thermonuclear weapons cannot be exaggerated. They are designed to destroy entire cities and regions, indiscriminately killing civilians and military.

702 A Chinese Dong Feng 5B ICBM (5 megaton warhead, 15,000km- range) surface explosion in lower Manhattan, NY, NY would produce a fireball that vaporizes a 2.39 km. radius (18km2).

703 In a 3.05km radius, a 29.2km2 area would receive 500 rem of radiation, sickening 100% with 50%-90% mortality occurring within hours to weeks given inadequate medical attention.

704 In a 3.72km radius, a 43.5km2 area suffers heavy damage from a 20-pound per square inch (psi) air blast that destroys even well-built cement structures; near 100% fatality rate.

705 In a 7.83km radius, a 193km2 area suffers a 5psi air blast that produces moderate damage with residential buildings collapsing, near universal injuries, and widespread fatalities.

706 In a 20.1km radius, a 1270km2 area receives a 1 psi air blast with light damage that sends glass flying into buildings after the earlier flash draws people to windows.

[1] Note: This hypothetical detonation is based on the Nukemap by Alex Wallerstein hosted by the College of Arts and Letters of Stevens Institute of Technology. On this interactive website, you can run hypothetical aerial or ground detonations with different warheads on your hometown or anywhere in the world, to capture the absolute horror that awaits humanity given the existence of nuclear weapons.

707 In a 21.3km radius, a 1430km2 area receives thermal radiation that causes third-degree burns on unprotected skin, heavy scarring, loss of functionality, and even amputation.

708 There are 9.5 million people within 20 kilometers of ground zero: 2.7 million die (28.4%), 2.6 million injured (27.3%), for a total of 5.3 million dead or injured (55.7%).

709 A 15kph SSW wind takes fatal 1000 rads radiation to Bayonne, New Jersey, and 100 rads with mild radiation sickness and reduced resistance to infection as far as Washington D.C.

710 Long-term effects of radiation include lymphatic cancer, carcinoma, bone marrow, and blood platelet damage, female infertility, and chromosome damage leading to birth defects.

EXTINCTION THREATS:

Nuclear Exchange Scenarios

711 The devastating effect of a thermonuclear attack on a region is exponentially increased by use of multiple independently targetable reentry vehicles (MIRVs).

712 Scenarios for total interchanges USA/Russian Federation are for between 1800 & 3000 warheads on MIRVs. Multiple targets in close proximity (e.g. NE U.S) lead to nuclear carpet bombing.

713 There are 230 U.S. Trident SLBM and 450 silo-based Minuteman. Russia has 264 ground-based missiles with hypersonic Avangard and 124 R-30 Bulava SLBM; additionally bombers.

714 The U.S. is developing low-yield (<20 kiloton, Hiroshima 15kt) tactical nuclear weapons and Prompt Global Strike (PGS) for attacks with missiles anywhere in world in one hour.

715 Next generation Russian weapons include the mach 20, sharp turn, vertical attack, practically unstoppable Avangard; also stealth, nuclear, limitless range drone, H-bomb torpedo.

716 Threats include high-altitude nuclear explosions for Electromagnetic Pulses (EMPs) to destroy electronics; anti-radiation missiles to destroy radar; cyberattacks on defenses.

717 Cybernetic attacks would reduce early warning, response, recovery capabilities. Attacks on nuclear power plants would add to radioactivity.

718 With half the population dead or injured and others exposed to immune system debilitating radiation, all human systems would fail as well as many natural ones.

719 Health, water, electricity, communications, transport, uncontaminated food, and public security would all fail amidst epidemics, famine, anarchy, and the end of our civilization.

720 Radioactive dust would cover the world's skies for at least a decade-long nuclear winter that impedes agriculture and disrupts the land and sea food chains, accelerating extinctions.

EXTINCTION THREATS:

Nuclear War Triggers

721 A nuclear war does not necessarily begin in the center, it can also be on the periphery. A rapid spiral of events out of control fueled by mistrust and ideology can lead to war.

722 Potential wars are Pakistan and India, Middle East Region (Israel, Saudi Arabia, Turkey, Iran, Syria, Lebanon, Iraq, Yemen); Ukraine-Russia; Baltic States-Russia; South China Sea.

723 From the very beginning of the nuclear age, we have not understood the consequences of nuclear weapons use. If we did, nuclear arms would immediately be abolished.

724 There has been a learning process, there are fewer nuclear accidents now, but resistance to nuclear weapons has declined. Missing are leaders like Einstein, Russell, and Pauling.

725 The large demonstrations of the 1950s and 1960s successfully contributed to the Nuclear Test Ban Treaty of 1963 and the Nuclear Non-Proliferation Treaty (NPT) of 1970.

726 The nuclear disarmament of the NPT is a dead letter in the context of the trillion-dollar modernization of nuclear weapons to make them more usable in "tactical settings."

727 Once the nuclear war genie is out of the bottle, it is hard to conceive of there not being escalations in use as well as the accumulation of areas contaminated for decades.

728 Low-yield nuclear weapons don't fall within MAD and hence rehabilitates the possibility of first strikes, but how does a side know what an incoming missile carries?

729 Rehabilitating first strike rehabilitates the possibility of preemptive strikes. With tactical nuclear weapons in use, every country will need at least a missile defense.

730 There was a nuclear war scare at the beginning of the Korean War & again in the 1962 Cuban Missile Crisis. Now people have grown accustomed to the threat despite the new arms race.

EXTINCTION THREATS:

Experimental, Energy, and Medical Nuclear Accidents

731 Nuclear ignorance leading to incidents, accidents, contamination, and deaths have characterized the nuclear age with regard to experiments, nuclear energy and medicine.

732 Two scientists died in Los Alamos, Harry Daghnian (1945) and Louis Slotin (1946), working on the "Demon Reactor." Cecil Kelley died in 1959 working with plutonium at Oak Ridge.

733 The Three Mile Island nuclear power plant accident (1979) spewed 2.8+k RU of radiation on 25,000 people in an 8km. radius who have above average rates of cancer and leukemia.

734 The Chernobyl #4 reactor exploded 26 April 1986, with radiation 30,000 x lethal level, killing 28 immediately and 19 subsequently. One million exposed to iodine fallout.

735 Chernobyl victims: 134-plus cases of radiation sickness and thyroid cancer cases in children; 120,000 displaced. Area of contamination (2600 km2); grade 7 nuclear disaster.

736 A tank went critical at the Toaimuri plant and killed Hiroshi Couchi (1999). The Fukushima Dai Ichi Nuclear Power Plant was hit by 9.0 earthquake and 15-foot tsunami (2011).

737 In the Grade 7 disaster at Fukushima, multiple systems failed because of earthquake damage, tsunami flooding, lack of electricity, and refrigeration, leading to explosions, fires, and meltdowns.

738 Radioactivity released into the air, sea (11.5k tons in 2016), and land. Fukushima damage will last 40-plus years. A 20km. exclusion area, 300k people evacuated, 160,000 permanently.

739 U.S. has 99 nuclear plants (99.8GWs, 17.7% of electricity), Russia 36 (26.1GWs, 17.1%), China 36 (31.3GWs, 3.6%), South Korea 25(23.GWs, 30.3%), France 58(53.1GW, 72.3%).

740 Nuclear medicine accidents include: United States, radium watch face painting "Radium Girls" 1904, Spain 1990 (11 dead), Panamá 2000 (28), Canada 1985 (3), Costa Rica 1996 (117).

EXTINCTION THREATS:

Nuclear Tests, Contaminated Sites, Accidents

741 Nuclear testing, contaminated sites, air/submarine accidents, systems failures, political calculations, and miscalculations and human failings have happened and will happen.

742 The 1954 test Castle Bravo (15mtns) was 2.5 times the calculated yield and spread radioactivity on Marshall Islands and the Japanese fishing boat Daigo Fukuyru Maru (1 crewman died).

743 The U.S. Nevada Test site had 928 tests, sent radiation to Utah, Idaho, Montana. U.S. Army and U.S. Marines competed on who could put their troops closest to explosions.

744 Semipalátinsk in Kazakhstan was Soviet test site (456 tests) until an uncontrolled explosion contaminated the site making it unusable (70km2); Mailuu-Suu (Kyrgzstan) unusable.

745 Sellafield in Cumbria, UK, was a bomb factory until an undetected fire spread radioactive contamination over 30km2. It daily dumps 8m liters of radioactive waste into sea.

746 Hanford, Washington, has 70% of the nuclear waste of the U.S., including 200m liters of liquid waste, 700m kilos of solid waste, and 500km2 of contaminated underground water.

747 The Somalian coast and the Eastern Mediterranean are nuclear waste dumps used by mafias from Italy and Somalia where the 2004 tsunami washed up barrels of uranium and cadmium.

748 Airborne accidents: Mediterranean (1956), Atlantic City (1957), Savannah (1958), Washington (1959). South Carolina (1958), Almería, Spain (1966), Greenland (1968).

749 Submarines: USS Thresher, Cape Cod 1963; USS Scorpion Azores 1968; Soviet K-129 Hawaii 1968; Soviet November Class, Biscay 1970; Russian Kursk Barents 2000; Chinese Ming 2003.

750 Lost warheads pose environmental risks but also threats from criminals/terrorists who sought weapons with the breakup of the Soviet Union, including suitcase-sized mini bombs.

EXTINCTION THREATS:

Korean War, Yom Kippur, Losing Sides, 1983

751 After Hiroshima and Nagasaki, the U.S. had a short monopoly on nuclear weapons broken by the first Russian test (1949). In Korea the use of nuclear weapons was considered several times.

752 General Bradley made a proposal to use nuclear weapons when the UN was confined to the Busan pocket. Instead MacArthur staged the Incheon landing, dividing enemy forces.

753 President Truman ordered General MacArthur not to approach Yalu River for fear of Chinese entry that occurred as General Ridgeways' forces marched North. Chinese then pushed UN back.

754 MacArthur wanted to use nuclear bombs in Manchuria and put a Cobalt C radiation band across peninsula to preclude passage. Truman relieved MacArthur of his command because of differences.

755 Truman ordered 4 Mark 6 thermonuclear weapons assigned to the military in case Chinese/N. Korean troops advanced too far. Armistice reached at the 38th parallel 27 July 1953.

756 In the Yom Kippur War (1978), Golda Meir authorized nuclear strikes on Egypt and Syria if troops advanced too far. After Soviet warnings to Israel, the U.S. went on high DEFCON 3 alert.

757 Korea and Yom Kippur indicate losing sides are most prone to use nuclear arms. It is not so much that they give an advantage, but are great equalizers for losing side.

758 Pakistan's first-use doctrine is congruent with losing side propensity. Indian Army could cut Pakistan in two south of Faisalabad (Lahore-Karachi railway and reach Afghanistan).

759 In 1983 the Soviets were convinced the West planned to decapitate their leadership and large NATO maneuvers in E. Europe seen as preparations; both sides on high alert.

760 In September, 1983, the Soviet Early Warning system indicated an attack, as did backup, but only five missiles in air led Lt. Col. Stanislav Petrov to label it a false alert.

EXTINCTION THREATS:

Military Industrial Complex, Bay Of Pigs

761 Ex-Supreme Allied Commander Europe and ex-President of the United States Dwight D. Eisenhower in January 1961 warned of the threat to democracy of the Military Industrial Complex.

762 Eisenhower in his Farewell Address stated that both a large standing army and a permanent armaments industry were new in the post-war, accumulating enormous unbridled power.

763 Eisenhower: "We must guard against the acquisition of unwarranted influence.... The potential for the disastrous rise of misplaced power exists and will persist."

764 Curtis "Bombs Away" LeMay, chairman of the Joint Chiefs of Staff, made the proposal to the new President in 1961 of a preemptive nuclear strike on the Soviet Union in 12-1963.

765 When Kennedy asked about why the December, 1963 date, he was told that it was the moment the USA would have a maximum advantage in warheads.

766 Kennedy walked out of the meeting and commented to Defense Avisor McGeorge Bundy, "How can we call ourselves human beings with this kind of proposal?"

767 Kennedy biographer Arthur Schlesinger sought to confirm meeting with Bundy who answered, "You mean the meeting in which the generals wanted to blow up the world."

768 Eisenhower turned over power in 1961 along with what would become the CIAs Bay of Pigs invasion by the 2503 Brigade of Cuban counterrevolutionaries controlled by the CIA.

769 CIA Plan included popular uprisings, military desertions, and destroying the Cuban air force on the ground; required landing site surprise and plausible deniability; none happened.

770 Cuban air force dispersed, outgunned B26s, strafed landing crafts and sites, landing forces trapped on beach and in swampy area, 1100 captured, U.S. role in all media: complete failure.

EXTINCTION THREATS:

Bay of Pigs Aftermath

771 The CIA did not really believe the 1500 strong Brigade could defeat 25,000 soldiers, 200,000 militia, and 9000 armed police; or that the obsolete B/26s the Cuban T-33 jets.

772 CIA and Brigade miscalculated U.S would intervene to avoid failure, but Kennedy maintained his position of no direct U.S. intervention, not even air support when Brigade trapped.

773 The CIA felt that they were being made scapegoats for mission that failed because of political restrictions imposed by President, and Brigade felt betrayed by President and CIA.

774 Kennedy felt ill served by CIA, which tried to entrap him to intervene, and by the Joint Chiefs of Staff that ill-advised him. He took steps to bring them under his control.

775 Allan Dulles (Director), USAF General Charles Cabell (Deputy Director), and Russell Bissell (Deputy Director Planning, covert operations) all dismissed before end of year.

776 Taylor Commission enquiry: separate intelligence from covert operations; CIA only for smaller ops; supervision CIA by Robert Kennedy and Taylor, members Foreign Intel. Board.

777 General Maxwell Taylor, close Kennedy aide, had more disagreements than not with Joint Chiefs; in September, 1962, named chairman of Joint Chiefs of Staff.

778 After the Bay of Pigs, expansion of economic blockade and covert Operation Mongoose: Castro assassination attempts, commandos, sabotage, psyops, fake news, media war.

779 A U.S. recon flight over Cuba 16-10-1962 photographs a Soviet Missile Base under construction, on 22-10 U.S. imposes naval blockade on Cuba; on 26-10 USSR announced departure of missiles.

780 The U.S. ended the naval blockade 20 November; resolved by U.S. pledge not to invade Cuba and retiring missiles in Turkey, as well as actual Soviet withdrawal of missiles from Cuba.

EXTINCTION THREATS:

October, 1962 Operational Close Calls in High Alert

781 The Cuban Missile Crisis was the closest to a nuclear war ordered by political leaders, but in an age of ICBMs the operational level can also make war and peace decisions.

782 Field officer decisions with high levels of tension like a DEFCON 3 alert led to close calls in the 25-28 October 1962 period, especially the 27th, the most dangerous day in history.

783 On October 25, 1962, a Duluth Air Force Facility guard fired at an intruder climbing the perimeter & activated an alarm that scrambled nuclear armed F-106As at Volk AFB, WI.

784 The jets were scrambled at Volk AFB because in DEFCON 3 (2 levels from nuclear war), there are no drills. When the intruder turned out to be a bear, the scramble was halted.

785 On October 27, 1962, a U-2 reconnaissance plane was shot down over Cuba, raising tensions to maximum level. Another U-2 in the Bering Sea area flew in Soviet airspace 90 minutes.

786 Six MiGs were sent to shoot down the intruder, and NORAD sent F-102s Delta Daggers armed with nuclear Falcon air-to-air missiles to protect it and avoid losing a second U-2 that day.

787 McNamara in the middle of this situation declared, "This means we are at war with the Soviet Union." Kennedy exclaimed, "There is always one son of a bitch that doesn't get it."

788 The Bering Sea MiGs didn't reach the U-2 that ran out of fuel, and the F-102s escorted it as it glided to Alaska. If they had met, there would have been a nuclear air battle.

789 *USS Beale* detected a submarine off Cuba & dropped training depth charges. The Soviet B59 submarine captain ordered a nuclear torpedo aimed at aircraft carrier *USS Randolff.*

790 The order to fire required the approval of all three senior officers. Second-in-command Vasili Arkapov disagreed, convinced the captain to surface and request orders from his HQ.

EXTINCTION THREATS:

Tapes, Switch, Chip, Moon, Reflecting Clouds

791 On October 23, 1962, at 03:00 the Moorestown, N.J., air defense radar showed incoming missiles from Cuba due to strike near Tampa in 6 minutes. Training tape had been loaded.

792 Training tape depicting a full Soviet attack was mistakenly placed in a NORAD computer (1979), scrambling planes before lack of satellite confirmation cancelled alert.

793 The Ballistic Missile Early Warning System went out in 1961, attack presumed; all systems used same relay station with burnt-out motor.

794 A rising moon over Thule, Greenland (1960), registered as an attack, disallowed as Khrushchev in New York. Soviet radar mistook reflected light off top of clouds as missiles (1983).

795 The summary meters of incoming ICBMs and SLBMs started reading 20 missiles, then 200 missiles, then 2. When being repaired, it malfunctioned again because of a faulty chip.

796 In 2010 the U.S. launch center lost contact with 50 ICBMs for an hour. Danger is presumption of attack on communications and early warning as prelude to attack.

797 On October 24, 1962, a Soviet satellite was launched and exploded. U.S. mistook it for a missile attack, according to Bernard Lovell of Jodrell Bank Observatory in 1968.

798 An unannounced Titan missile test from Florida to the South Pacific alarmed Moorestown Radar, and another from Vandenburg AFB certainly alarmed the Soviets on 26 October 1962.

799 On January, 25, 1995, Soviet early warning detected a missile launch from Spitzbergen with 5-minute flight time to Moscow. Norway had announced, Soviet Early Warning not privy.

800 On 26 September 1983, the month KAL 007 shot down, Soviet radar showed 5 incoming ICBMs. Col. Stanisav Petrov reasoned U.S. attack would be 100s. Cause: reflections on top of clouds.

EXTINCTION THREATS:

Falling Bolt, Faulty Invoice, Solar Flares, Power Outage, Kurile Test, Mechanics, Errant Airmen, NATO Nuclear War Games

801 In maintenance of Titan II missile silo in Damascus, Arkansas (September 19, 1980), a 9-pound socket fell 70 feet, caused fuel leak, explosion, airman dead, warhead blown off.

802 Six-armed cruise missiles without being identified as such flown from Minnesota to Louisiana (29-30 August 1974) and left unguarded for 36 hours, not checked to see if armed.

803 General radar failures interpreted to be hostile and prelude to attack, jets mobilized, ordered to stand down after radar problem attributed to solar flares (May 23, 1967).

804 A power failure alarm system registered red (nuclear blast) in two cities, mobilizing Command Center of Office of Emergency Planning but not military (November 9, 1965).

805 In a test 4 submarine missiles near Kurile Islands (March 15, 1980), one with trajectory appearing headed to U.S. Threat assessment mechanism activated, determined false alarm.

806 During Arab-Israeli War (October 24, 1973), U.S. went on DEFCON 3 to warn USSR; mechanics at Kinchole AFB, Michigan accidentally activated alarm that scrambled B-52s.

807 Drug use of ecstasy and speed at Warren (2016) and Malmstron (2014) AFBs; 19 lost launch authority; more than half of 181 airmen involved in cheating on proficiency tests.

808 Realistic NATO war games with new codes, radio silences, consultations with Heads of State, culminated in a simulated DEFCON 1 (war with Soviet Union) on 11 September 1983.

809 The Soviets feared real attack underway under guise of exercises, mobilized when intercepted a NATO message, “the US missiles have already been launched against the USSR.”

810 Robert Gates, CIA: “We might have been on the brink of war and not even known it.” Murphy’s law is in full force in the nuclear age with an amazing amount of things that can fail.

EXTINCTION THREATS:

High-Level Decision Making

811 When the Norwegian Spitzbergen missile was fired, President Boris Yeltsin was informed and immediately declared it to be a false alarm. All of this within 5-minute response time.

812 Both Harry Truman in the Korean War and Golda Meir in the Yom Kippur War ordered the use of nuclear weapons if an advancing enemy advanced "too far," avoidance of defeat.

813 The *USS Pueblo* was captured by North Korea 40 km. off coast with 1 petty officer killed and 82 held captive for 11 months. Ship is still tourist attraction in Pyongyang.

814 The world awaited the U.S. response, which never came for fear of reigniting war on the Korean peninsula while U.S. already overextended in Vietnam.

815 On 15 April 1969, North Korea shot down a U.S. EC-131 electronic recon plane 90 kilometers off the North Korean coast, killing all 33 crew members. World awaited response.

816 President Nixon had to cope with this complex, stressful situation, and he did so by getting drunk, and in that state he ordered a tactical nuclear attack on North Korea.

817 National Security Advisor Henry Kissinger complied with the President's order but also advised the Joint Chiefs of Staff to ignore it until confirmation after President awoke in morning.

818 USAF pilot Bruce Charles at Kunsan AFB, South Korea, on standby to undertake a nuclear attack on a North Korean AFB as his part of SOIP attack plan, next day ordered to stand down.

819 At end of Watergate crisis in August, 1974, President Nixon was clinically depressed, emotionally unstable, and drinking heavily. This illustrates the danger of nuclear power concentrated in one person.

820 On 1 August 1974, Secretary of Defense Schlesinger instructed Joint Chiefs to channel any strategic order from the President through him, such as a nuclear launch.

EXTINCTION THREATS:

Kennedy Policy Shift and Magnicide to Provoke War

821 After Bay of Pigs and Cuban Missile Crisis, Kennedy opened hotline to Moscow, back channel with Khrushchev, sent emissaries to Castro, planned withdrawal from Vietnam.

822 Kennedy's motivation for this policy shift was clearly delineated in a speech at American University on 10 June 1962 in which he called for world peace and disarmament.

823 Kennedy: "Total war makes no sense in an age when great powers can maintain large and relatively invulnerable nuclear forces and refuse to surrender without resort to those forces."

824 Kennedy: "It makes no sense in an age when a single nuclear weapon contains almost ten times the explosive force delivered by all of the allied air forces in WWII."

825 Kennedy: "Our primary long-range interest in Geneva, however, is general and complete disarmament—designed to take place by stages."

826 Kennedy: "(disarmament) permitting parallel political developments to build the new institutions of peace which would take the place of arms."

827 On 23 November 1963, one month before the date of the 1961 Joint Chief of Staff proposal for a pre-emptive strike, Kennedy is assassinated in a plot pointing to USSR and Cuba.

828 The alleged sole shooter Lee Harvey Oswald was an ex-marine who lived in the USSR and returned to the U.S. with his Russian wife.

829 Oswald arrested in New Orleans in "Hands Off Cuba" leaflet distribution scuffle (August 3, 1963), allegedly visited Soviet Embassy in Mexico City, and allegedly called a KGB agent in Mexico City shortly before the assassination.

830 After the assassination, the FBI Director phoned President Lyndon B. Johnson to inform him that the photos and voice of the visitor/caller to the Soviet Embassy in Mexico were not Oswald.

EXTINCTION THREATS:

Evidence of Kennedy Assassination Conspiracy

831 President Johnson convincing Senator Russell to serve on Commission investigating Dallas stated, "You don't know what I am up against, the voice in the KGB call wasn't Oswald's."

832 The Warren Commission found Oswald to be the sole shooter, but there is evidence of at least one shot from the front, indicating a second shooter and hence a conspiracy.

833 Dallas Parkland Hospital physicians Dr. Robert McClelland and Dr. Charles Cranshaw have maintained that there were front entry and a large, gaping back exit wound.

834 Lt. Cmdr. William Pitzer USN, Bethesda Naval Hospital Audio/Visual Director, filmed rear exit wound, film seen by USN Chief Petty Officer Dennis David, Hospital Corpsman.

835 Pitzer found dead by gunshot in 1966 at this Bethesda studio before retirement and TV network job offers. NCIS: suicide. Family: murder. JFK corpse film and slides: missing.

836 Lt. Col. Vern Marvin, retired Green Beret and CIA hitman, called Pitzer's widow (1993) and said he was asked to kill Pitzer at Bethesda before he retired, refused because in USA.

837 Airman Robert Vincent, Jr., on standby at Andrews AFB to fly to Colorado, told of a plane heading west, boarded empty plane, as did two pilots, landed flood plain Dallas.

838 When plane landed in Dallas two men boarded and flight continued to a locked down Roswell AFB. Later Vincent assigned to CIA at Area 51, Nevada, with double pay and secrecy oath.

839 After retirement and protected by Kennedy Assassination Records Act (1992), Vincent publicly told his story, which lends credence to official agency involvement in magnicide.

840 Johnson (November 24, 1963): "I am not going to lose Vietnam. I am not going to be the president who saw Southeast Asia go the way China went"; in 1968, 500,000 U.S. troops.

EXTINCTION THREATS:

Policy Plots

841 The Kennedy Assassination condenses nuclear risk given the attempt to induce nuclear war (e.g., false KGB call). Possible avoidance of peace policies by all means, including Vietnam.

842 The avoidance of all-out nuclear war in the Cuban missile crisis depended on two men. There are no institutions or fail-safe mechanisms to save us from our greatest risk.

843 In high-tension alert situations, mistakes increase and field decisions can start a war: submarine captain, jet bomber pilot, a radar officer can all make war or peace decisions.

844 Marshall Schulman, Sr., State Department Advisor wrote: "False alerts of this kind are not a rare occurrence. There is a complacency about handling them that is disturbing."

845 That most of the incidents come from the United States and USSR is more a matter of reporting than reality. More incidents occurred in earlier years because of declassification.

846 Actually the situation is worse now than in earlier years because of advances in hacking. They could affect energy supply, early warning, launch orders, control of weapons.

847 Not necessary to have nuclear weapons to control them. Besides cyberwarfare, opportunities for cybercrime and blackmail further complicate hacking in a nuclear age.

848 As a result of policies, out-of-control conflicts, losing side refusing defeat, or mistakes or accidents, nuclear arms do not defend us, but rather threaten to destroy us.

849 A total of 4000 cocked warheads on the Planet is not rational, they are the maximum irrationality of the warped logic of the industrial-military complex that advocates trillion-dollar modernization.

850 Tactical weapons modernization increases possibility of use, first strikes, and preemptive strikes. Either we destroy nuclear weapons or they will very surely destroy us.

EXTINCTION THREATS:

Anthropogenic, Chemical, and Biological Weapons

851 Chemicals were first used in poison bullets limited by Strasburg Agreement (1675), Brussels Convention (1874), and Hague Conferences (1899 and 1907); Protocol of Geneva (1925).

852 In World War I gases were first used by the Germans at the Second Battle of Yepes (22-04-1915), who attacked French, Canadians, and Algerians. In war 50,965tns used by both sides.

853 Total of 1.3 million soldiers killed WWI (90,000 by gas). Protocol of Geneva (1925) forbad use but not production of chemical weapons: Used by Spain in Morocco, Britain in Russia and Mesopotamia, Italy in Abyssinia, and Japan in China.

854 In the Vietnam War the U.S. applied napalm and 20 million tons of Agent Orange, causing prostate and thyroid cancers. The Soviets/Russians created nerve agents: Novichok, Sarin, Tabun.

855 In the 1980s there was use of chemical weapons in Iran-Iraq War and in Halabja with 3,200-plus killed. In Syrian war the opposition undertook false flag propaganda gas attacks.

856 The Convention Against Chemical Weapons (CAQ) signed 13 January 1993, Paris, 192 adherents, destroyed 90% of weapons; enforcement by OPAQ, formed in the Hague (1997).

857 Biological weapons were used by Mongols (bubonic plague), Assyrians (tularemia), and British colonialists (Australia, North America, small pox). Japan used them in China occupation.

858 U.S. biological weapons program (1944-1969) weaponized 7 bio-agents, Russia 11. U.S. and British programs (from WWI-1972) discredited by experiments on non-consenting populations.

859 The Biological Weapons Convention (BWC, 1972) supplements Geneva Protocol (1925) and obliges 181 states to prohibit production of biological and toxin weapons but no verification.

860 Nanotechnologies, new materials, self-replicating factories, artificial intelligence, and drone delivery combined with chemical and biological weapons increase current threat.

EXTINCTION THREATS:

Anthropogenic, Climate Change, Ozone Layer

861 Apart from the nuclear, the greatest man-made threat is climate change with CO2 going from 280 parts per million (ppm) in 1750 to 350ppm (1988) and 413 ppm (2020).

862 In 2012 the World Bank declared "We Can and Must Avoid a 4 Degree World," the business as usual result, warning of catastrophic consequences if public policy is not changed.

863 The Paris Accord of 2015 set goals of limiting the world temperature rise to 2.0°C or preferably 1.5° C, but the 2015 Paris voluntary national commitments lead to 2.7°C to 3.7°C

864 The IPCC report 1.5°C (2018) requires emission reductions of -45% by 2030 and building a sustainable, zero-emission society by 2050 to be able to limit global warming to 1.5°C.

865 Increased ambition must come from the 10 major emitters within 72% of emissions. The 100 smallest emitters mostly in South represent 3%; they can best contribute by reforestation.

866 Ozone layer stabilized after world efforts to implement the Montreal Protocol and should recover by 2070 despite agricultural fertilization emitting depleting Nitrous Oxide (N2O).

867 Nuclear/chemical/biological warfare threats and biotech-engineered pandemics for military weapons/ bioterrorism/ biocriminal blackmail or from bio errors are top threats.

868 Artificial Intelligence (AI) threats include rogue autonomous robots ("tech singularity"), robot soldiers, "automated battlefields" with robots making decisions.

869 Nanotechnology threats include molecular weapons manufactured in 1-kilo nano-factories, autonomous self-propagation until saturation, AI, and biotech threats.

870 Except for ozone depletion, if nothing changes, none of the anthropogenic threats is even under partial control. We ourselves are creating our greatest risks.

EXTINCTION:

Extinction Risk Assessment

871 Risks are higher for anthropogenic risks, including nuclear mass destruction; given our propensity for war and violence, if we have them we will eventually use them.

872 Climate change is ongoing and already contributing, along with habitat loss, to the Holocene Event, through increasing temperatures, sea levels, extreme weather, droughts, floods.

873 Nuclear arms and self-replicating, decision-making robots can be prohibited, AI technologies can be controlled if there is political will, as of yet lacking.

874 Natural threats, bolides, and supervolcanoes, occurring at longer intervals are real. Historically they have happened and they will happen again, only when?

875 A 1km asteroid impacts earth every 440,000 years on average. A 10km asteroid impact 65 mya in Chicxulub created a 180 x 20 km crater and the K-Pg dinosaur extinction event.

876 A supernova explosion within 33 lyrs takes place on average every 240 million years. One in Centauro 250 mya led to an extinction event that reduced sea life by 50%.

877 Supervolcanic eruptions occur on average every 55,000 years. The Toba explosion close call for humans occurred 70 tya, without anything even remotely similar afterward.

878 Epidemiological risks include completely new viruses, mutations, and changed forms of transmission in existing ones, as well as increased resistance to medicines.

879 Our 200 tyr journey as an *H. Sapiens* and 4 mys as *Genus Homo* has evaded extinction, but we must also avoid "survivors bias" that somehow this will continue.

880 We live in a cosmic shooting gallery of bolides, solar radiation, & gamma rays, on a molten bomb of a planet, attacked by pathogens, unable to control our science & technology: nuclear weapons, climate change, artificial intelligence.

EXTINCTION:
Priorities

881 What should be the priority in time, energy, thought, and resources to prepare for risks that are very low probability but have catastrophic consequences.

882 Low probability/catastrophic consequence risks should neither receive "no priority" nor receive "high priority." Neither of these positions is rational.

883 Also irrational is "No or Low Priority" for risks that are a clear and present danger as well as have transversal relations with multiple threats.

884 Climate change consists of global warming, droughts, and floods, an increase in frequency/intensity of extreme weather events, and negative changes in agricultural production.

885 Climate change represents habitat loss, migratory disruptions, increase in invasive species, augmented water, and food scarcity and food chain disruption.

886 Climate change results in decreased food and water security and increased epidemiological risks; loss of biodiversity; local/in the wild/total extinctions.

887 We can mitigate climate change, adapt to it, and compensate/replace losses and damages, as compared with cosmological/geological risks where chance prevails.

888 We invented nuclear weapons and we can abolish them. However, this requires, as does climate change, overcoming hegemonic economic, political, and military domination.

889 Temperature is a critical variable that has conditioned the decoupling of the forces and elements of the Universe in the Great Expansion and stellar fusion.

890 We emerged as a species within a given temperature range, and there are parameters beyond which our thermal regulatory systems start to break down, +50°C. and -50°C temperatures.

EXTINCTION:

Resilience

891 A survival strategy option combines objectives/goals, policies/planning with contingencies, emergency planning, for synergies between short-, medium-, long-term programs/actions.

892 The unit of analysis and action is the species *homo sapiens sapiens*. There are no partial, local solutions to truly global threats but only division of labor,

893 The objective is to increase societal resilience through preparation and mitigation where possible, loss and damage response/recovery, adaptation for future.

894 Civilization increases our resilience but to a point. Beyond that point largely autarkic family and community producers fare better.

895 High levels of specialized social organization have systems vulnerability. Systems failure in any systems leads to breakdowns in the rest.

896 Systems failure is particularly acute in relation to electrical energy, transportation, citizen security. Without these three, anarchy replaces civilization.

897 The policies/plans must address root causes of anthropogenic climate change, food/water insecurity; deforestation/habitat/biodiversity losses/extinctions.

898 The long-term sacrifice of the lives of future generations and other species on the altar of immediate, transient material gratification must be overcome.

899 The solution entails overcoming the unsustainable model of production/consumption/lifestyles and its political/media/military/police power structure.

900 The solution is a Survival Social Movement to construct a low- or no-carbon, climate-resilient, recyclable, much more equal, sustainable society, free of nuclear weapons.

Epistemology of Existence and Extinction

The Theory of Everything is Information

901 The Theory of Everything is information. The Singularity was the cosmic stem cell that had all the information to structure the Universe according to its lifecycle.

902 The Great Expansion began with pure energy at the maximum temperature, velocity, and curvature of space-time. As it cooled and slowed it expanded.

903 At different temperatures, velocities, and volumes, the forces unfolded, gravity, strong force, weak force, electromagnetism along with the bosons that carry them.

904 Matter was formed when the conditions, temperature/velocity/density information, as well as the forces, allowed combinations from collisions and accretion.

905 Structures have information on the possibilities and impossibilities for internal transformation of a given entity at a determinate stage in its lifecycle.

906 Conditions determine the outcomes of the possibilities at a given stage of the lifecycle. Conditions are information, as is the impact of conditions on structures.

907 Combinations of structures and processes form entities (galaxies, stars) of different types whose lifecycles, by type, are information.

908 Spaces for the possibility of chance events are structured by information and their probabilities are information. Change is structured but not deterministic for individual outcomes.

909 Access to existential information is through genetic/sensory/perceptive/cognitive processes that are and produce information.

910 Information is the theory of existence and the theory of knowledge. The nexus between ontology and epistemology and everything is information.

Sensory Perception Information

911 We perceive/process data/produce information because the Universe consists of conditioned, structured, not random, transformations.

912 The transformations are structured by, and the conditions consist of, information. We access information through sensory perception.

913 Sensory perception is a physiological process whose data are processed by other physiological processes, one for each sense. They all require interpretation to become perception.

914 We are bombarded by abundant, continuous sensory data that require differential attention and valoration. We have innate filters and create filters to prioritize.

915 Processing is accomplished by sensory frameworks that sort/evaluate/interpret the data and convert it into sensory information.

916 Sensory interpretation is most complex. It involves continuous streaming of perception and interpretation, comparisons and inferences.

917 Comparisons and inferences involve analysis and synthesis, deduction and induction and synergies with these processes in other senses.

918 All of these sensual perception processing centers draw upon libraries of stored sensory material for their comparisons/inferences.

919 We receive sensory information in the form of smells, tastes, touch, images, sounds at the sensory level; processed also for other levels (e.g., emotions).

920 The sensory information is processed and combined, as per relevance and priorities, to produce cognitive information that is based on sensory induction and cognitive deduction.

Levels of Perception, Information

921 Reactions to perceptions can be **physiological/neurological**, for example, sexual stimulation or survival reactions, including fight or flight responses.

922 At a second level, basically linked to our DNA inheritance, there is the level of **impulses**, closely related to that of the **sub-conscious.**

923 At the impulse/sub-conscious level, the reaction to sensory information or an entity can be that of phobia, neutrality, philia based on associations with previous stored experiences.

924 Reflexes are part of impulses and the physiological systems, as is "the zone" of slow-motion intense focus, the "automatic pilot" that is a halfway house between consciousness and the subconscience.

925 Stream of consciousness and dreams are part of the subconscious and are related to memory, the sensory library filing, and retrieval system.

926 Most movement is impulse. We don't think "I want to stand up" and then stand. We start to stand up and consciousness has a split-second veto.

927 A fourth level of **emotions** combines all the other levels to produce or not feelings that are positive/good, negative/bad, by strength.

928 A fifth level of **conscious thought** combines the other levels with a world view, rationality, logic, for interpretation, synthesis, and analysis.

929 All of these levels consist of both genetically transmitted firmware and software that is programmed in our life experience.

930 Physiology, impulses, the sub-conscious, emotions, and conscious thought are interrelated in complex ways in the production of knowledge.

The Five Components of Knowledge Production

931 The five components of knowledge production are description, explanation (includes prediction), evaluation, preference, and obligation.

932 **Description** consists of constructing a conceptual taxonomy of the characteristics relevant to a given entity (structure/process) and classifying a given field in terms of it.

933 Description also consists of applying the taxonomy to an entity or group of entities to produce descriptions in its terms. Reality is reconstructed in terms of the taxonomic framework.

934 A new field of learning or knowledge is taxonomical. It is necessary to be able to describe before being able to produce explanation/prediction.

935 **Explanation** consists of fulfilling the terms of a mode of explanation to relevant satisfaction (see Modes of Explanation below).

936 An explanation functions in the past or present, while prediction is an explanation of future structures or processes; it is an explanation in future tense.

937 **Evaluation** is the valoration of an entity or group and/or explanation in terms of one/more values: larger/smaller, better/worse, others.

938 **Preference** is evaluation in terms of the personal values of the evaluator: like/dislike (personal) or approve/disapprove (ethical).

939 Ethical preference is borderline with ethical obligation, adding an imperative "everyone should" element, but still voluntary. Peer pressure or professional ethics can make noncompliance difficult.

940 Moral **obligation** implies the imperative "everyone must," which is obligatory whether enshrined in a code of conduct, legal norm, or religious doctrine.

Three Modes of Knowledge Production

941 The levels of knowledge production are common sense; intuitive, intermediate, and ideological; and formal and scientific.

942 The **common sense and intuitive** production of knowledge has low consciousness of the knowledge-production process and low methodological rigor.

943 Basic rationalities, logics, worldviews transmitted by culture structure much of common-sense knowledge. Most of our daily activity is based on common sense knowledge.

944 If one asks young children in the Orient and Occident about a wooden block, a rubber ball, and a metal triangle, **cultural conditioning** is apparent.

945 Children in the Orient answer "What is This?" with wood, rubber, and metal, whereas in the Occident they answer a block, a ball, and a triangle, substance vs. form.

946 A vast amount of our daily knowledge that is crucial for our daily activity is common sense, and it also underlies and is present in the other levels.

947 **Intermediate and ideological** knowledge has greater self-consciousness of the process of knowledge formation and more explicit logic and methodology.

948 **Formal and scientific** knowledge has logical and methodological rules and rigor, including explicit theoretical and epistemological frameworks.

949 Any regression of a line of scientific discourse eventually leads to ideological and finally common sense "core assumptions"; our thought is unified.

950 In scientific thought, the three interrelated modes of knowledge production organize contents of the five components in epistemological/theoretical frameworks.

Cultural Conditioning of Epistemological Frameworks

951 Oriental and Occidental thought have fundamentally different assumptions as the substance/form difference at even very young ages demonstrates.

952 Oriental thought emphasizes transience and cycles as in the Hindu cycle of reincarnation or Buddhist satori or nirvana as deep understanding of transience.

953 Yin/Yang and endless subsidiary concepts construct a unity of opposites within an integrated, holistic worldview that stresses equilibria and harmony, inner peace.

954 Occidental thought has emphasized eternity and stability. The Judeo-Christian-Islamic tradition and monotheism stress eternal truths; promise eternal life.

955 The ideal is separated from the material, and the ideal created the material and its development. Humans are created in the image of the ideal.

956 Revelation and reason are part of the ideal and the source of our knowledge. The ideal-human relation and eternal life place humans outside/above Nature.

957 Materialism in Western thought is a reaction to Idealism. It separates the ideal and the material and posits that the material determines the ideal.

958 Both idealism and materialism separate the ideal and the material but differ on which is the active subject and which is the passive object.

959 Western thought tends to be causal and lineal, whereas Oriental thought favors **EQUILIBRA**, harmony. The difference is apparent in the basic precepts of Western and Oriental medicine.

960 **Sensory** perception is our window on the Universe and the World and all thought and theories, the ideal, are constructed from perception and induction/deduction, synthesis/analysis.

EPISTEMOLOGICAL FRAMEWORKS:

Empiricism

961 Formal and scientific knowledge is produced with explicit epistemological frameworks containing positions on how we produce and justify knowledge.

962 The epistemological frameworks are empiricism, logical positivism, structuralism, pragmatism, and dialectical materialism.

963 Empiricism posits knowledge production by the direct/neutral observation of reality and justifies it by demonstrating its correspondence to reality.

964 The end result of the process of producing knowledge is "laws," considered universal and immutable given the conditions they contain are satisfied.

965 Theory and practice are unrelated in the process of producing knowledge. The practice of rigorous observation is all that is needed to produce knowledge.

966 Sense experience translates mechanically to perceptions and these automatically to knowledge. "The facts speak for themselves," without interpretation.

967 The "real," "natural" connections between phenomena do not require cognitive elaboration. Interpretative theory is projecting upon nature human ends.

968 Observation is the measure of knowledge. Empiricism is an absolutistic epistemology in which there is one reality and one true knowledge that corresponds to it.

969 Theory opens the door for interpretations and relativism with values forming the criteria for interpretative choices. Science should be value free.

970 Empiricism is widespread in common-sense knowledge, but it entered into crisis with the rise of experimental psychology in science, which debunked "the facts speak for themselves."

EPISTEMOLOGICAL FRAMEWORKS:

Logical Positivism

971 The same person can have different perceptions of the same ambiguous drawing (for example, the young/old lady drawing) by a shift in interpretation, gestalt.

972 The production of scientific knowledge for the logical positivists begins with the postulation of an hypothesis of any conceivable origin.

973 The testing of a hypothesis takes place by the confrontation between a theoretical framework and observations of reality. There are different criteria for acceptance.

974 A theoretical framework has defined concepts, theoretical presuppositions, directive questions, and a mode of explanation.

975 A mode of explanation stipulates the logic and rules that determine what are to be considered acceptable answers to directive questions.

976 Reality consists of relevant "facts," "events" that are identified by the concepts, theoretical presuppositions, and directive questions of a framework.

977 Theoretical frameworks define reality but are also based on prior research and empirical theory. Observation reduces the possible without exhausting it.

978 The practice of observation is informed by theory and theory informed by prior observation. Reality, however, is one thing and theory another.

979 Within the possible reduced by eliminating the manifestly absurd by observation, there is room for different interpretations of reality.

980 Different theoretical frameworks can lead to different interpretations of even the same observations and scientific revolutions because of paradigm shifts.

EPISTEMOLOGICAL FRAMEWORKS:

Justification In Logical Positivism

981 Logical positivism differs from empiricism in the role of theory in knowledge production, introducing relativism and eliminating "correspondence to reality."

982 The results of processing a hypothesis by observation guided by a theoretical framework are "scientific theories" and discarded hypotheses.

983 "Scientific theories" are corrigible and tentative. There can be differing scientific theories to explain the same phenomenon. Science is guided by its own values.

984 The logical positivist criteria of justification is method. It is the method by which knowledge is produced that justifies knowledge itself, not "correspondence."

985 Two logical positivist positions related to the products of knowledge production are the probabilistic and the falsification theories of verification.

986 The probabilistic theory of verification assigns probabilities to different theories explaining a phenomenon. High-probability theories have the best claim to be knowledge.

987 Given relativism, there is caution in assigning probabilities; empiricists describe "A causes X," but logical positivism posits "A conditions X."

988 The Theory of Falsification identified with Karl Popper maintains that theories can only be falsified with all theories, not eliminated as possibilities.

989 The task of science in the Theory of Falsification is to falsify theories versus assigning them probabilities in the Probabilistic Theory.

990 The move from empiricism to logical positivism in formal/ scientific knowledge is from absolutism to relativism by the introduction of theory in production.

EPISTEMOLOGICAL FRAMEWORKS:

Structuralism

991 In structuralism knowledge is produced by transforming abstract knowledge (Generality I) into concrete knowledge (Generality III).

992 The transformation is by a mode of production of knowledge (Generality II). Generality I comprises elaborated abstractions that may come from any source.

993 Generality II, the mode of production of knowledge, consists of a structured body of concepts that define problems by the confrontation between facts/concepts.

994 Concrete thought, Generality III, produces the "effect of knowledge." It is tentative and corrigible and should not be confused with concrete reality.

995 The criterion for the justification of knowledge is method. The proof of the pudding is not in the eating but rather the mechanism that identifies it is pudding.

996 Louis Althusser: "It is not the end product that characterizes and classifies reflection, but rather the mode or method of reflection."

997 The object of knowledge is theoretical practice not to be confused with a real object. The ideal (knowledge) and the material (practice) are separate.

998 Real objects exist independently of knowledge, although they can only be defined by knowledge. Theory/practice consists of theoretical practice.

999 There is no pure theory, absolute vision, nor is there totally material practice. There is no practice in general, but specific practices.

1000 The various practices are distinguished by the mode of production with which they transform their types of objects, i.e., raw materials, into end products.

EPISTEMOLOGICAL FRAMEWORKS:

Pragmatism

1001 There are two human reactions to uncertainty and peril: change self in thought (religion, philosophy), and change objective reality through action (science).

1002 In Occidental thought, weak, fearful humans sought eternity and certainty by the divorce of the ideal/material and theory/practice for security.

1003 The process of secularization was merely a shift from an imaginative myth-laden emotional style to a more rational discourse governed by canons of logic.

1004 Salvation by rituals was replaced by salvation by abstract reason. The net effect of both religion and philosophy was the same, theory/practice divorce.

1005 From Greek civilization they distinguished a higher realm of fixed reality and a lower realm of changing things that only offers precarious probability.

1006 For pragmatism nature is neither rational nor irrational but rather intelligible and understandable by humans who are entirely within nature.

1007 The production of knowledge begins with practical problems. Science is an activity. It is a mode of directed, practical doing guided by values and ideas.

1008 Actions undertaken to maximize desired values are guided by ideas. Science is based on action, the operationalization of ideas.

1009 The justification of knowledge is judged by the consequences of an operation, if the action fulfills the predictions of the directive idea.

1010 The action should maximize the appropriate values and resolve the problematic situation; then it is justified as knowledge.

EPISTEMOLOGICAL FRAMEWORKS:

Dialectical Materialism

1011 The question for dialectical materialism is not how abstract humans produce knowledge but how socio-historic humans in society evolved knowledge.

1012 Nature transforms humans, and humans transform nature in the process of obtaining their subsistence for survival. Knowledge is produced by this interaction.

1013 Humans are molded by their natural and social environments, but active, selective, critical, and creative human intelligence also has the capacity to mold nature to an extent.

1014 Our intelligence and action mold our natural and social environments. This occurs through the union of theory and practice, praxis, which produces knowledge.

1015 The epistemological implications of praxis are that between the material and the ideal there are no set subjects or objects of reality.

1016 The objective natural and social environments interact with sense-based human intelligence in praxis, the dialectical union of theory and practice.

1017 Human needs, socio-historically defined in concrete contexts, are the point of departure in the process of knowledge production and the point of final arrival.

1018 Knowledge is produced by individuals in society by their activity in their material conditions of life, those found and those produced by their activity.

1019 Our knowledge production, communication, storage, retrieval, and dissemination are social activities that have a history, including cultural and technological innovations.

1020 Socio-historic human needs is the justification of knowledge, value-based social utility, driving praxis, the combination of theory and practice to solve our problems.

Values, Theory, and Practice

1021 Values are value free in empiricism. Values do not constitute knowledge but rather are ideological and interfere with knowledge production.

1022 Empiricism rejects theory that is simply our projecting our values on reality. Interpretation is unnecessary inasmuch as the facts speak for themselves.

1023 Logical positivism considers that the only values appropriate in science are the values of the scientific method itself. Other values contaminate science.

1024 Scientific theory interprets data in terms of the concepts and rules of the theoretical framework utilized, but it in turn is empirically based.

1025 Structuralism accepts values within theoretical practice, which is not real, objective practice but which contains its own subject and object.

1026 In structuralism abstract theory is processed by a theoretical mode of production to produce concrete theory that produces the "effect of knowledge."

1027 Pragmatism considers that values and theories are of utility if they contribute to solving the practical human problems whose solution produces knowledge.

1028 Dialectical materialism is socio-historically based and hence value based in relation to the resolution of historical situations of human needs.

1029 The criterion of justification in dialectical materialism is social utility in a specific socio-historical context in human social evolution. It is social utility that justifies knowledge.

1030 Dialectical materialism posits the dialectical union of theory and practice, praxis, as the logical framework for the generation of knowledge.

The Three Modes of Explanation

1031 The three modes of explanation are causal-lineal, teleological, and dialectical, which integrates elements of the first two. These are the modes accepted by science.

1032 The causal-lineal mode posits A causes X or A conditions X with X being explained (dependent variable) and A the explaining (independent variable).

1033 The simplest form of causal-lineal is bivariate; with two or more independent variables, the explanation is multivariate, also with the possibility of intermediate variables.

1034 The ontological supposition of causal-lineal is that all phenomenon can be explained by other phenomenon that cause or condition them. This begins simple but soon becomes complex.

1035 The teleological mode posits that A fulfills or realizes X with X being an objective or end (teleos) or motivation of the actors involved. We often explain by what motivated.

1036 Empiricism only accepts causal-lineal explanations based on causes, while patterns of causes are immutable laws of nature given the conditions stipulated are met.

1037 Logical positivism only accepts the causal-lineal mode but posits "conditions" rather than "causes" because of the relativism of probabilistic/falsification proof.

1038 Pragmatism accepts all modes of explanation of utility, but they must be operationalized, while its own logic is teleological, to fulfill human ends.

1039 Dialectical materialism posits internal transformations that are structured (teleological with limits) and externally conditioned (causal-lineal).

1040 Other modes of explanation are possible, but the ones presented are those employed in formal, scientific knowledge production historically and currently.

Ontology, Epistemology, Ethics

1041 The ontology of existence and extinction exists independently from human intelligence, but our consciousness is totally within the ontology.

1042 Our human consciousness does not condition the transformations of the Universe, but we have affected and can affect some structures and conditions of Mother Earth.

1043 Inasmuch as the expansion of the Universe is structured and conditioned, with a pinch of chance, not random or chaotic, we can know its logic and evolution.

1044 The nexus between ontology and epistemology is the theory of everything, information. Expansion of the Universe and evolution of Earth is based on information.

1045 Our sensory experience is data that interpretation transforms into perceptions that with cognitive interpretation can become knowledge.

1046 Our access to the structures and processes of the Universe is based on sensory induction and cognitive deduction in a dialectical unity of opposites.

1047 Our concepts are inexact but can function with "family resemblance," and our logic and methodologies have limitations, but we do produce useful knowledge.

1048 The difference between common sense, ideological, and scientific knowledge is quantitative in terms of explicitness and rigor, not a difference in kind.

1049 In practice we combine the components, levels of knowledge, and modes of explanation in continuous knowledge production of dynamic, changing reality.

1050 Our knowledge is socio-historic and value oriented. The combination of knowledge and values organized in an ethical system is wisdom, what we most lack and need.

ETHICS OF EXISTENCE AND EXTINCTION

Meaning

1051 The structures and processes of the Universe are structured and conditioned, within its logic, and intelligible, but without meaning.

1052 Meaning begins with life in that the meaning of life for all living things is to live it and survive. This is in our genetic firmware and in our consciousness.

1053 Knowledge production is purposive, social activity and hence value driven, and the evaluation of the social utility of results is value based.

1054 Purposive, social activity is intentional and directional, guided by values derived from our historic social and individual objectives.

1055 It is not that our knowledge is objective and our values subjective, but rather our knowledge and values have the same origin, praxis.

1056 We construct purpose in our species and our individual lives through our values, ethical and moral systems, and our observance or not of them.

1057 Intelligent consciousness leads to existential questions and cultural/religious myths of origin. It leads to philosophy and science searching for answers.

1058 The construction of purpose from values, ethical and moral systems is part of species, cultural and individual identity and evaluations.

1059 Humans **evalu**ate applying values. Some evaluatory criteria are very widespread, against genocide, killing, robbery, rape. Others are more culturally specific and not universal.

1060 Our generation will be evaluated on if we reverse processes threatening future generations, including global warming and the maximum crime, speciescides.

Inherent Values

1061 We recognize transmitted instinct in other species, but it's difficult to recognize it in our own genetic inheritance. We still try to distance ourselves from other animal species.

1062 Inherent values are transmitted in the DNA firmware that operates at the physiological, impulse, emotional/subconscious levels.

1063 **Life and survival** are the prime inherent values. All of the organs and systems contribute to homeostatic equilibrium for survival.

1064 Our organism has resources similar to servomechanisms to proactively maintain EQUILIBRA for survival. For example, mechanisms to raise or lower our temperature.

1065 Instantaneous responses for survival include fight or flight, corticol and adrenalin activating heart beat increase, blood vessel and pupil dilation, tunnel vision.

1066 **Life** itself is the highest value for all living things. **Survival** is the continuation of life within the full extent possible of the lifecycle.

1067 Physiological, impulse, emotions and subconscious, sex response, and sex drive are species and individual DNA transmission expressions of the survival value.

1068 Derived from life and survival are safety (freedom from danger), wellness (freedom from illness) and well-being (freedom from want), hydration, nutrition, shelter.

1069 Much of our historical and current time and effort are devoted to realizing safety, wellness, and well-being; only recently there is time to pursue other values.

1070 Life and survival are species values inherited from the evolution of life on Earth. They are not relative. They are absolute as they protect all we have, space-time on Earth.

Social Values

1071 Inherent values are related to Human-Nature relations conditioning survival. Human-Human relations generate social values for survival and fulfillment.

1072 Social relations were also almost exclusively for survival when human vulnerability had been the greatest and still are today to a great extent.

1073 Increasing social differentiation and complexity make survival life support less visible but no less real. It becomes more efficient but also has its own vulnerability.

1074 Advanced civilization generally provides stronger safety, wellness, and well-being than highly independent family producers unless there is systemic breakdown.

1075 Human-Human relations are at the family, clan, community, locality, national, regional, and species levels, which have evolved across our evolution/history.

1076 The prime social values are **love** and **solidarity** with collateral values of affection, sympathy/empathy, caring, sharing, common to both love and solidarity.

1077 The long gestational period, helpless offspring upon birth, defenseless early childhood, and extended learning period require intensive, prolonged parenting.

1078 None of us would survive infancy or early childhood without love and solidarity, affection, sympathy, empathy, caring, and sharing from those who parent us.

1079 These values are also transmitted to us by our parental mentors as well as by our interactions with them and clan care and mentoring in some societies.

1080 *Homo sapiens sapiens* is a social species, and, like all social species, we are dependent on each other for life and survival from birth. In civilization we are dependent on social systems that organize an entire society.

Constructed Values

1081 Our social practice seeks safety, wellness, and well-being for survival, but we also strive for fulfillment, constructing needs/values in the process.

1082 Social needs and values have been increasingly constructed as society became progressively more differentiated and complex.

1083 With one notable exception, we are not a particularly well-endowed species. Many species can naturally hear, smell, touch, see, run, swim, climb, fight better.

1084 Some species also have highly developed radar and sonar. Some navigate around the world with great precision. Many are better at fight and/or flight.

1085 Some species are armed with impressive defensive mechanisms. We are not the biggest or the strongest, nor do we have subtle camouflage, but we are the cleverest.

1086 We are able to transform our natural and social environments intentionally, but not necessarily in the sense of doing so wisely.

1087 Our intelligence is our comparative advantage, and it is social intelligence. We have developed the capacity to learn, create, and accumulate knowledge socially.

1088 In our social praxis, we intentionally transform our natural and social realities and articulate, communicate, store, and retrieve knowledge through language.

1089 **Learning** and **knowledge** have been and are vital for life and survival, but also for constructed needs and values for achievement, fulfillment, and self-realization.

1090 **Learning** and **knowledge** themselves can provide fulfillment/self-realization by production, creativity, arts, discovery, exploration, sports, altruism, others.

Ethical and Moral Values

1091 Ethics are value configurations and rules that everyone "should" follow; morals add the imperative of "must" follow backed by religious and/or social and/or legal sanctions.

1092 Core values are: life and survival; love and solidarity; learning and knowledge; culture and institutions; equality, equity, justice; peace; dignity, human rights, human security.

1093 Conflict is inherent in all social life and may be channelled peacefully or violently. Violence is human self-destruction or destruction of others.

1094 The values **equality, equity, and justice** reduce exclusion and domination, creating conditions for a culture of conflict resolution, strengthening institutions.

1095 **Peace** was desirable before nuclear weapons; now it is a moral imperative. **Dignity** (freedom from humiliation), **human rights** (freedom from oppression), **human security** (freedom from danger) elevate values to ethics/morals.

1096 Values, attitudes, beliefs, world views, customs, behaviour patterns, and material artefacts, including tools, utensils and art, are the **culture** received and constructed.

1097 Formal and informal rules, including laws, incentive structures, and national/organizational **culture**, form **institutions** that structure our interdependence.

1098 Culture and institutions based on ethical and moral values are fundamental in civilization, generational, group, individual identity, and our evaluations.

1099 Cultures and institutions also have their lifecycles and require energy to renew, enhance, develop. They can face threats of slow onset degradation or cataclysmic disaster.

1100 Our generation will be judged on if we can overcome processes now threatening ourselves, future generations, and other species.

Ethical and Moral Evaluation

1101 Our generation will be evaluated on our ability to overcome the threats of continued weapons of mass destruction, culture of violence, and full-spectrum dominance.

1102 We must also overcome climate change, global warming, environmental degradation, and the maximum crime against nature, species extinctions, including our own species suicide.

1103 The knowledge we require for survival is of the philosophy, economics, politics, policies, and practices that lead to survival, informed by science to the maximum extent.

1104 In 2017, in *Science* (March) and *Nature* (June), top climate scientists J.J. Schnellhuber and J. Rockström and climate leader C. Figueres and 68 more identified a pathway to 1.5°C.

1105 Science informs us that we need to half CO2(e) emissions every decade to 2050: beginning with 40 Gtn in 2020, reduce to 24 Gtn in2030, 14 Gtn in 2040, and 5 Gtn in 2050.

1106 Halving emissions would give us a 66% chance to limit the world average temperature increase to 2°C and a 50% chance for 1.5°C by 2100, thus achieving the targets of the Paris Accord.

1107 The IPCC report on 1.5°C. demands -45% emissions by 2030, a net zero-emissions sustainable society by 2050. If we fail to limit to 1.5°C, even 2°C entails extinction of coral.

1108 If the world goes to 3°C, we will have 4°C to 6°C in the tropics, deserts, and Arctic regions. Even now at 1° above the pre-industrial level, we have catastrophic storms and Niños.

1109 At 3°C the U.N. Sustainable Development goals will be compromised. There will be less water, food, and health and more poverty and death if climate change is not controlled.

1110 The immoral inter-generational injustice we will bestow on our children and grandchildren: our hedonistic pleasures and luxuries now for their future of increasing misery and death.

Survival Movement

1111 Are we capable of forming a Survival Movement for a low-carbon, climate-resistant and resilient, circular, sustainable society with far greater equality and no nuclear weapons?

1112 Our ability to struggle for survival in the face of hegemonic structures is vital for ourselves and future generations. Actually nothing else matters in the face of the very real threat of climatic or nuclear extinction.

1113 A scientist asked about the probability that aliens would visit Earth searching for intelligent life replied his greatest fear was that they had arrived, searched, and left.

1114 On environmental degradation, including climate change, as well as nuclear weapons, we have hardly exhibited intelligence as a species, given self-destructive behavior

1115 In our second and probably last chance, Mother Earth can supply all we need but definitely not our greed, but we must overcome our "business as usual" self-destructive behavior.

1116 We have the science and technology, but much of our knowledge is ignored or devoted to continue endless, limitless, mindless production, consumption and capital accumulation.

1117 A misuse of science and technology that continually grows in our time is the alienation of "scientificism," which holds that science and technology will always find a way to save us.

1118 We have the financial resources, but they require radical reallocation from anti-values. More than one trillion a year is spent on militaries that could destroy us rather than on climate action that could save us.

1119 The survival movement must mobilize social and political will based on moral strength and spiritual wherewithal combined with ultimate collective self-interest and altruism.

1120 We are in the Universe and on planet Earth, a unique individual of our species within life and existence, our greatest opportunity and our greatest responsibility.

UN Photo / Kim Haughton

DR. PAUL OQUIST is Minister-Private Secretary for National Policy for the President of the Republic of Nicaragua, Commandant Daniel Ortega Saavedra. He has been a Nicaraguan citizen since 1981.

Dr. Oquist is a former Member of the Board of Directors of the Green Climate Fund, and he was elected by the developing countries in the Green Climate Fund to represent them as Co-Chair of the Board of Directors in 2018 and now remains an Advisor. Dr. Oquist lobbied for the organization of the Green Climate Fund in COP-16 in Cancun, served on the Transition Committee (2010–2011) that produced a proposal, while Nicaragua was selected for the final negotiation in Durban by the G-77+China with the United States that represented the developed country constituency.

Dr. Oquist is a third-term member of the Standing Committee of Finance (SCF) of the UNCCC. He represented the SCF in the Interim Directorate of the Warsaw Mechanism of Losses and Damages.

He was a member for two consecutive terms (2010–2017) of the United Nations Economic and Social Council (ECOSOC) Committee of Experts on Public Administration (CEPA).

Dr. Oquist has been Nicaragua's Head of Delegation in every Conference of the Parties of the Framework Convention on Climate Change since the last part of COP-15 in Copenhagen. In COP-21, in Paris, Dr. Oquist communicated the Nicaraguan position not to sign the agreement and to demand a greater level of ambition from large emitter countries in accordance with their historical responsibilities, in order to limit global warming to 1.5°C in this century. By COP-23 in 2017 in Bonn, the need for greater ambition was a consensus position of most countries, and Nicaragua signed and ratified the Paris Accord. President Daniel Ortega Saavedra in his message to the UN General Assembly in 2015 maintained that countries with climate change losses and damages have a right to claim indemnization and compensation from the largest historical emitters (1880–to date) in proportion to their level of responsibility. The Nicaraguan delegation has maintained this as the only position that is both scientific and just.

Dr. Oquist was also the spokesperson for the ALBA countries at COP-15 in Copenhagen in their defense of the multilateral negotiating process in the face of the attempt to impose a parallel document initiated by the largest emitters without multilateral negotiations that take into account developing countries. The one-sided document was successfully blocked.

In recent COPs Nicaragua has stressed the urgency of the current situation based on the latest scientific information available and given the gravity of recent catastrophic climate events. The Report on the 1.5°C. on which Nicaragua and Bolivia insisted in COP-15 was assigned to IPCC and presented in October, 2018. The findings are that to limit climate change to 1.5°C. average world temperature in this century, it is necessary to reduce emissions 45% by 2030 and achieve a zero-emission, sustainable society by 2050.

Dr. Oquist was Senior Adviser to the President of the 63rd General Assembly of the United Nations, Father Miguel D'Escoto Brockmann. In that capacity he was Chief Negotiator for the Second Financing for Development Meeting held in Doha, Qatar, in November, 2008, which surprisingly approved a high-level United Nations meeting on the "Financial and Economic Crisis and its Impact on Development" that Dr. Oquist coordinated on behalf of Father D'Escoto in June, 2009. He was also liaison with the group of high-level economists who prepared an Expert Paper for the Conference under the direction of Nobel-winning Professor Joseph Stiglitz of Columbia University.

Dr. Oquist was Regional Expert on Governance for the Asia Division of the United Nations Development Program (UNDP) based in Islamabad, Pakistan (1998–2006). He was Senior Technical Adviser for the UNDP Management Program in Mongolia (1993–1998) and Director of UNDP Public Administration Projects in Quito, Ecuador (1975–1979). In addition, Dr. Oquist was a consultant for UNDP and UNICEF at the highest level of government in Colombia, Ecuador, Kyrgyzstan, Vietnam, and the Philippines, where he undertook eight policy assessments on "Peace and Development in the Southern Philippines" for the Multi-Donor Support Group between 1999 and 2006.

In 2019, Dr. Oquist represented Nicaragua in: 49th Session of the Intergovernmental Panel on Climate Change (Kyoto, Japan); United Nations Climate Action Summit (New York, U.S.A.); Conference of the Parties of the United Nations Convention to Combat Desertification (New Delhi, India); 2019 Forum of the Standing

Committee on Finance on "Climate Finance and Sustainable Cities" (Beirut, Lebanon); High Forest Cover and Low Deforestation Conference on Climate Finance Mobilization (Paramaribo, Surinam); Fourth Session of the UN Environment Assembly (UNEA-4) (Nairobi, Kenya); Petersberg Climate Dialogue X (Berlin, Germany); XVII Summit of the Heads of State and Government of the Non Aligned Movement (Baku, Azerbaijan); CELAC V Think Tanks Forum (Beijing, China); Our Oceans Conference (Oslo, Norway); CEPAL Sustainable Development Conference (Mexico City, Mexico); in the Italy – Latin America Caribbean Conference (Rome, Italy); Consejo Político ALBA-TCP (Havana, Cuba) and the 10th Bosphorus Summit (Istanbul, Turkey).

In 2018 Dr. Oquist represented Nicaragua with interventions on Nelson Mandela in the General Assembly and in defense of Venezuela in the Security Council of United Nations (New York, U.S.A.); in the Ministerial Conference on Nuclear Science and Technology (Vienna, Austria); and the Global Insurance Forum (Berlin, Germany), among numerous others.

In recent years he has represented Nicaragua in water summits (Hungary, 2013; Abu Dhabi, 2014); Protected Maritime Areas (Marseilles and Ajaccio, 2013) and Convention of Cartagena on protection of Caribbean (Cayenne, French Guyana, 2016); renewable energy (IRENA, Abu Dhabi; Government of Great Britain, London, U.K., 2015); and political and economic affairs (NO-AL, Bali, Indonesia, 2013). He has represented the President of the General Assembly in NO-AL, (Havana, 2009); Organization of African Unity, (Addis Ababa, 2008) and UNDP at the British Commonwealth Conference on Local Governance (Aberdeen, Scotland, U.K, 2005), Conference of Asian Political Parties (Bangkok, Thailand, 2002) and Millennium Conferences (Seoul and Tokyo, 2000); among others.

BACKGROUND

Dr. Oquist holds a Ph.D. (1976) and an M.A. (1967) in Political Science from the University of California, Berkeley, while his bachelor in Political Science was from U.C.L.A. (1966). He did a postgraduate course at the Latin American School of Political Science and Public Administration of the Latin American Faculty of Social Sciences (FLACSO), Santiago, Chile. His doctoral dissertation "Violence, Conflict and Politics in Colombia" was published in Spanish and English (Institute of Colombian Studies, 1978, and Banco Popular, several editions) (Academic Press, 1980). He was Professor of Political Science at the University of the Andes in Bogotá, Colombia (1970–1975). He has also taught in FLACSO (Santiago and Quito) and the Latin American Post-Graduate School of Economics (ESCOLATINA) of the University of Chile.

Dr. Oquist was born in Oak Park, Illinois, USA, whose primary school system was field-visit intensive in the museum-rich Chicago area. He attended Roosevelt Junior High School and Compton High School in Compton, California, while he graduated from Belmont High School (Summer, 1961) in Los Angeles, California, where he was Senior Class President, Class Valedictorian and voted "Most Likely to Succeed."

Dr. Oquist has lived in Oak Park, Illinois; Russellville, Arkansas (mother's hometown); Compton, Los Angeles, Oakland, Berkeley, and Albany, California; Santiago, Chile; Bogotá, Colombia; Quito, Ecuador; New York City, New York; Ulan Bataar, Mongolia; Islamabad, Pakistan, and at home in Managua, Nicaragua. His father's hometown was Svensbyn, Piteå, Norrbotten, Sweden. He has been married to Pilar Aguilar Matamoro for over 50 years, whose hometown is el Carrizal, Ciudad Darío, Matagalpa, Nicaragua.

Made in the USA
Monee, IL
02 May 2026

49411616R00154